I0488645

Total Nitrogen and Suspended-Sediment Loads and Identification of Suspended-Sediment Sources in the Laurel Hill Creek Watershed, Somerset County, Pennsylvania, Water Years 2010–11

By Ronald A. Sloto, Allen C. Gellis, and Daniel G. Galeone

Prepared in cooperation with the
Somerset County Conservation District

Scientific Investigations Report 2012–5250

U.S. Department of the Interior
U.S. Geological Survey

U.S. Department of the Interior
KEN SALAZAR, Secretary

U.S. Geological Survey
Marcia K. McNutt, Director

U.S. Geological Survey, Reston, Virginia: 2012

For more information on the USGS—the Federal source for science about the Earth, its natural and living resources, natural hazards, and the environment, visit http://www.usgs.gov or call 1–888–ASK–USGS.

For an overview of USGS information products, including maps, imagery, and publications, visit http://www.usgs.gov/pubprod

To order this and other USGS information products, visit http://store.usgs.gov

Any use of trade, product, or firm names is for descriptive purposes only and does not imply endorsement by the U.S. Government.

Although this report is in the public domain, permission must be secured from the individual copyright owners to reproduce any copyrighted materials contained within this report.

Suggested citation:
Sloto, R.A., Gellis, A.C., and Galeone, D.G., 2012, Total nitrogen and suspended-sediment loads and identification of suspended-sediment sources in the Laurel Hill Creek watershed, Somerset County, Pennsylvania, water years 2010–11: U.S. Geological Survey Scientific Investigations Report 2012–5250, 44 p.

Contents

Figures

Tables

Conversion Factors

Inch/Pound to SI

Multiply	By	To obtain
Length		
inch (in.)	25.4	millimeter (mm)
foot (ft)	0.3048	meter (m)
mile (mi)	1.609	kilometer (km)
Area		
square mile (mi^2)	2.590	square kilometer (km^2)
Volume		
gallon (gal)	3.785	liter (L)
cubic foot (ft^3)	0.02832	cubic meter (m^3)
Flow rate		
cubic foot per second (ft^3/s)	0.02832	cubic meter per second (m^3/s)
gallon per minute (gal/min)	0.06309	liter per second (L/s)
million gallons per day (Mgal/d)	0.04381	cubic meter per second (m^3/s)
Mass		
ton, short (2,000 lb)	0.9072	megagram (Mg)
ton per day (ton/d)	0.9072	megagram per day (Mg/d)
ton per day per square mile [(ton/d)/mi^2]	0.3503	megagram per day per square kilometer [(Mg/d)/km^2]
ton per year (ton/yr)	0.9072	megagram per year (Mg/yr)

SI to Inch/Pound

Multiply	By	To obtain
Length		
centimeter (cm)	0.3937	inch (in.)
meter (m)	3.281	foot (ft)

Temperature in degrees Celsius (°C) may be converted to degrees Fahrenheit (°F) as follows:

°F=(1.8×°C)+32

Temperature in degrees Fahrenheit (°F) may be converted to degrees Celsius (°C) as follows:

°C=(°F-32)/1.8

Vertical coordinate information is referenced to the North American Vertical Datum of 1988 (NAVD 88).

Horizontal coordinate information is referenced to the North American Datum of 1983 (NAD 83).

Elevation, as used in this report, refers to distance above the vertical datum.

Concentrations of chemical constituents in water are given either in milligrams per liter (mg/L) or micrograms per liter (µg/L).

Acronyms

FNU	Formazine nephelometric units
Q	Discharge
SSC	Suspended-sediment concentration
SSL	Suspended-sediment load
SSY	Suspended-sediment yield
T	Turbidity

Total Nitrogen and Suspended-Sediment Loads and Identification of Suspended-Sediment Sources in the Laurel Hill Creek Watershed, Somerset County, Pennsylvania, Water Years 2010–11

By Ronald A. Sloto, Allen C. Gellis, and Daniel G. Galeone

Abstract

Laurel Hill Creek is a watershed of 125 square miles located mostly in Somerset County, Pennsylvania, with small areas extending into Fayette and Westmoreland Counties. The upper part of the watershed is on the Pennsylvania Department of Environmental Protection 303(d) list of impaired streams because of siltation, nutrients, and low dissolved oxygen concentrations. The objectives of this study were to (1) estimate the annual sediment load, (2) estimate the annual nitrogen load, and (3) identify the major sources of fine-grained sediment using the sediment-fingerprinting approach. This study by the U.S. Geological Survey (USGS) was done in cooperation with the Somerset County Conservation District. Discharge, suspended-sediment, and nutrient data were collected at two streamflow-gaging stations—Laurel Hill Creek near Bakersville, Pa., (station 03079600) and Laurel Hill Creek at Ursina, Pa., (station 03080000)—and one ungaged stream site, Laurel Hill Creek below Laurel Hill Creek Lake at Trent (station 03079655).

Concentrations of nutrients generally were low. Concentrations of ammonia were less than 0.2 milligrams per liter (mg/L), and concentrations of phosphorus were less than 0.3 mg/L. Most concentrations of phosphorus were less than the detection limit of 0.02 mg/L. Most water samples had concentrations of nitrate plus nitrite less than 1.0 mg/L. At the Bakersville station, concentrations of total nitrogen ranged from 0.63 to 1.3 mg/L in base-flow samples and from 0.57 to 1.5 mg/L in storm composite samples. Median concentrations were 0.88 mg/L in base-flow samples and 1.2 mg/L in storm composite samples. At the Ursina station, concentrations of total nitrogen ranged from 0.25 to 0.92 mg/L in base-flow samples; the median concentration was 0.57 mg/L. The estimated total nitrogen load at the Bakersville station was 262 pounds (lb) for 11 months of the 2010 water year (November 2009 to September 2010) and 266 lb for the 2011 water year[1].

Most of the total nitrogen loading was from stormflows. The stormflow load accounted for 76.6 percent of the total load for the 2010 water year and 80.6 percent of the total load for the 2011 water year. The estimated monthly total nitrogen loads were higher during the winter and spring (December through May) than during the summer (June through August).

For the Bakersville station, the estimated suspended-sediment load (SSL) was 17,700 tons for 11 months of the 2010 water year (November 2009 to September 2010). The storm beginning January 24, 2010, provided 34.4 percent of the annual SSL, and the storm beginning March 10, 2010, provided 31.9 percent of the annual SSL. Together, these two winter storms provided 66 percent of the annual SSL for the 2010 water year. For the 2011 water year, the estimated annual SSL was 13,500 tons. For the 2011 water year, the SSLs were more evenly divided among storms than for the 2010 water year. Seven of 37 storms with the highest SSLs provided a total of 65.7 percent of the annual SSL for the 2011 water year; each storm provided from 4.6 to 12.3 percent of the annual SSL. The highest cumulative SSL for the 2010 and 2011 water years generally occurred during the late winter. Stormflows with the highest peak discharges generally carried the highest SSL.

The sediment-fingerprinting approach was used to quantify sources of fine-grained suspended sediment in the watershed draining to the Laurel Hill Creek near Bakersville streamflow-gaging station. Sediment source samples were collected from five source types: 20 from cropland, 9 from pasture, 18 from forested areas, 20 from unpaved roads, and 23 from streambanks. At the Bakersville station, 10 suspended-sediment samples were collected during 6 storms for sediment-source analysis. Thirty-five tracers from elemental analysis and 4 tracers from stable isotope analysis were used to fingerprint the source of sediment for the 10 storm samples. Statistical analysis determined that cropland and pasture could not be discriminated by the set of tracers and were combined into one source group—agriculture. Stepwise discriminant function analysis determined that 11 tracers best described the 4 sources. An "unmixing" model applied to the 11 tracers showed that agricultural land (cropland and pasture) was the

---

[1]A water year is the 12-month period from October 1 to September 30. It is designated by the year in which it ends.

major source of sediment, contributing an average of 53 percent of the sediment for the 10 storm samples. Streambanks, unpaved roads, and forest contributions for the 10 storm samples averaged 30, 17, and 0 percent, respectively. Agriculture was the major contributor of sediment during the highest sampled stormflows. The highest stormflows also produced the highest total nitrogen and suspended-sediment loads.

Introduction

Laurel Hill Creek, located mostly in Somerset County, Pennsylvania, drains a 125 square-mile (mi^2) area (fig. 1). Laurel Hill Creek is classified by the Pennsylvania Department of Environmental Protection (PaDEP) as a High Quality Coldwater Fishery with four Exceptional Value tributaries. The upper part of the Laurel Hill Creek watershed is on the PaDEP 303(d) list of impaired streams (Pennsylvania Department of Environmental Protection, 2012a) because of sedimentation (siltation), elevated concentrations of nutrients, and low dissolved oxygen concentrations (fig. 2). The impaired stream segments are all upstream from the streamflow-gaging station on Laurel Hill Creek near Bakersville, Pa. (station number 03079600). Sedimentation can lead to a decline in surface-water quality and biodiversity. Sediments fill the interstices of gravel and cobble stream bottoms, decreasing the spawning areas for many fish species and the habitat for macroinvertebrates, which serve as food for many fish species. Nitrogen concentrations elevated above background concentrations in surface water can decrease the pH by releasing hydrogen and ammonia ions. Acidification reduces the biodiversity of aquatic ecosystems. Elevated nitrogen concentrations can speed up eutrophication, increasing algae growth and lowering dissolved oxygen concentrations.

As part of Pennsylvania's State Water Plan (Pennsylvania Act 220), Critical Water Planning Areas (CWPA) are designated where water use exceeds water availability, and Critical Area Resource Plans (CARPs) are developed for each CWPA. The PaDEP Ohio Water Resources Regional Committee designated Laurel Hill Creek as a regional priority for designation as a CWPA. In Act 220, a CWPA is defined as a "significant hydrologic unit where existing or future water demands exceed or threaten to exceed the safe yield of available water resources" (Pennsylvania Department of Environmental Protection, 2006a). The average daily quantity of water withdrawn from the Laurel Hill Creek watershed was 2.27 million gallons per day (Mgal/d) in 2003 and 2.22 Mgal/d in 2009. These values are approximately 150 percent of the 7-day 10-year low flow ($Q_{7,10}$) value for the watershed, which was estimated to be 1.43 Mgal/d (Simko Consulting, Inc., 2011).

A Cold Water Heritage Partnership PL-566 grant was developed for Laurel Hill Creek to address sediment and nutrients from farms (U.S. Department of Agriculture, 1987). For the PL-566 grant, it was estimated that approximately 48,400 tons of sediment are delivered annually to the creek

from areas in the upper part of the watershed. It was also noted that the sedimentation severely affects native trout populations and that the drinking-water reservoir for the Somerset Water Authority was filling with sediment

The Somerset County Comprehensive Plan (Somerset County Planning Commission, 2006) recognizes that "The County has some of the steepest agricultural land in the state and as a result it is also highly prone to erosion. The eroded topsoil ends up in the streams and rivers as sediment, which negatively impacts water quality and the health of fish and wildlife." Impaired water quality from sediment is listed by the U.S. Department of Agriculture (1987) as one of three areas of concern in the Laurel Hill Creek watershed. Often it is the fine-grained part of the sediment (silts and clays) that is detrimental to the environment (Gellis and Walling, 2011; Larsen and others, 2010). In order to effectively reduce sediment, it is necessary to identify the major sources of fine-grained sediment. This information can then be used to determine the appropriate management practices to reduce the nutrient and suspended sediment loads. Load reductions are necessary in order to remove affected stream segments from the PaDEP 303(d) impaired stream list.

The objectives of this study in the Laurel Hill Creek watershed were to (1) estimate the annual nitrogen load, (2) estimate the annual suspended-sediment load, and (3) identify the major sources of fine-grained sediment using the sediment-fingerprinting approach. This study by the U.S. Geological Survey (USGS) was done in cooperation with the Somerset County Conservation District.

Purpose and Scope

This report presents discharge, nutrient, and suspended sediment data collected at three sites in the Laurel Hill Creek watershed—Laurel Hill Creek near Bakersville, Pa. (station 03079600), Laurel Hill Creek at Ursina, Pa. (station 03080000), and Laurel Hill Creek below Laurel Hill Creek Lake at Trent, Pa. (station 03079655)—during the 2009–11 water years. Water samples for nutrients and suspended sediment were collected during low-flow and stormflow events to characterize concentrations and loads. The report presents total nitrogen and suspended-sediment loads and yields for Laurel Hill Creek near Bakersville for the 2010 and 2011 water years. The important sources of fine-grained sediment in the Laurel Hill Creek watershed upstream from the Bakersville station were identified using the sediment-fingerprinting approach. Nitrogen and suspended-sediment loads in base flow or stormflow are shown in illustrations and listed in tables.

Study Area

Laurel Hill Creek is a 125-mi^2 watershed mostly in Somerset County, Pa., with small areas extending into Fayette and Westmoreland Counties (fig. 1). Laurel Hill Creek enters the Casselman River approximately 400 ft upstream from the

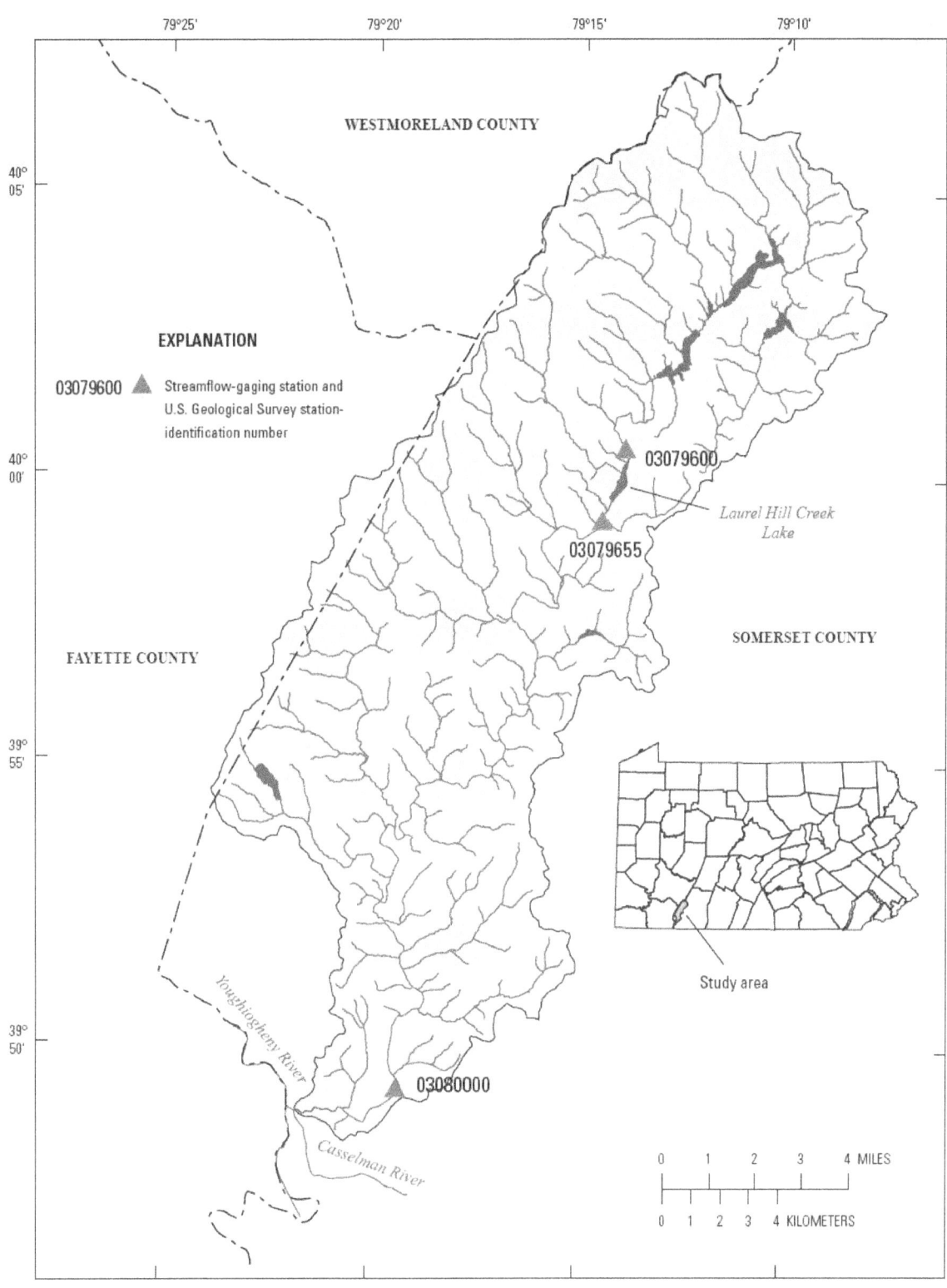

Figure 1. Location of the Laurel Hill Creek watershed and surface-water sampling sites, Somerset County, Pennsylvania.

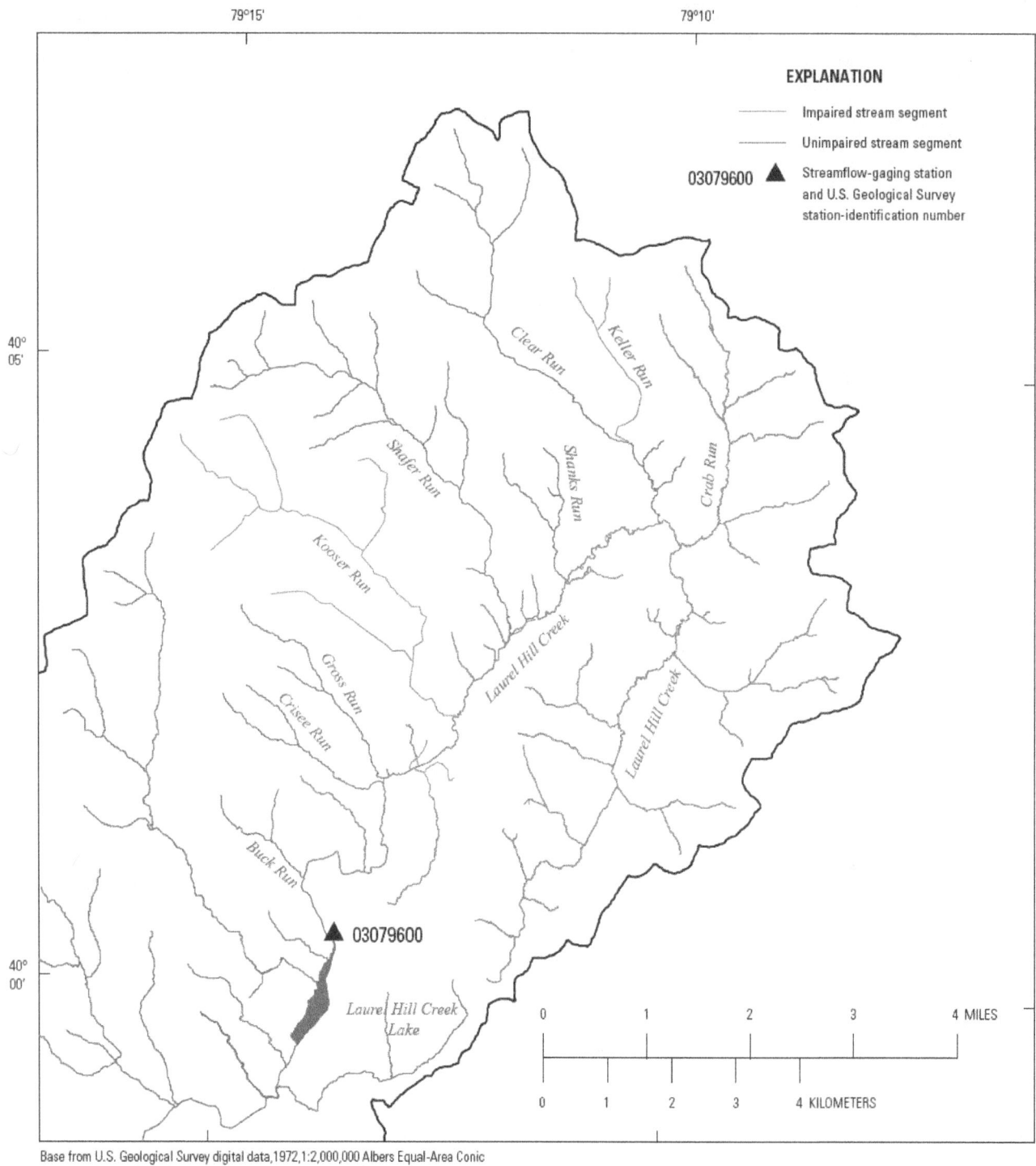

Base from U.S. Geological Survey digital data,1972,1:2,000,000 Albers Equal-Area Conic
Projection.Standard parallels 29°30'N, central meridian 75°00'W.

Figure 2.   Impaired stream segments in the Laurel Hill Creek watershed, Somerset County, Pennsylvania.

Youghiogheny River in the Borough of Confluence, Pa. This watershed has great recreational value because it contains three state parks (Laurel Hill, Kooser, and Laurel Ridge), Forbes State Forest, and State Game Lands 111. Several ski resorts are located on Laurel Hill in the western part of the watershed. Numerous vacation and second homes are located in and around the ski resorts.

The climate of the Laurel Hill Creek watershed is continental, but temperatures are more variable and precipitation occurs more frequently than in other parts of the State. Air temperatures show variation across the watershed as a result of orographic influences. Annual average minimum air temperatures range from 35.6 to 39.3°F (degrees Fahrenheit), and the annual average maximum air temperatures range from 53.5 to 61.5°F. On the basis of data from continuous recording stations in and around the watershed, the highest daily maximum and minimum temperatures occur in July and the lowest daily maximum and minimum temperatures occur in January. The daily maximum temperature ranges from about 75 to 85°F in July, and the daily minimum temperature ranges from about 15 to 16°F in January (National Oceanic and Atmospheric Administration, 2001).

Annual precipitation in the watershed varies spatially because of the mountainous terrain, leading to orographic influences on precipitation patterns (fig. 3). The 1971–2000 normal annual precipitation is 52.41 in. at Laurel Mountain and 42.38 in. at Somerset (National Oceanic and Atmospheric Administration, 2001). Precipitation is distributed fairly evenly throughout the year. Snowfall varies greatly across the watershed, both spatially and seasonally. Seven Springs Mountain Resort is situated at the top of Laurel Hill in the northwest corner of the watershed. Snowfall totals recorded by the resort from fall 2005 through spring 2010 indicate an annual average snowfall for the period of 135 in., with the minimum occurring during the 2008–09 season (98 in.) and the maximum occurring during the 2009–10 season (223 in.) (Jeff Alcorn, Seven Springs Mountain Resort, oral commun., 2010). The average annual snowfall was 54.9 in. for Confluence during 1971–2000. The snow totals for Seven Springs Mountain Resort and Confluence give a reasonable estimate of the snowfall within the watershed.

Physiography, Geology, and Land Use

The Laurel Hill Creek watershed lies within the Allegheny Mountain Section of the Appalachian Plateaus Physiographic Province. The Allegheny Mountain Section consists of broad, rounded ridges separated by broad valleys. The ridges decrease in elevation from south to north, and the ridges have no topographic expression at the northern end of the section. The ridges occur on the crests of anticlines that have been eroded to expose the resistant rocks that form the crests of the ridges. The southern parts of these ridges form the highest mountains in Pennsylvania. The valleys are broad, undulating surfaces with shallow to deep stream incision (Sevon, 2000). There is relatively large relief in the Laurel Hill Creek

watershed for Pennsylvania, with elevations ranging from 1,300 ft at the mouth of the watershed at Confluence, Pa., to approximately 2,990 ft on Laurel Hill at various locations along the ridge top (fig. 4).

Sedimentary rocks of Pennsylvanian and Mississippian ages are exposed in the Laurel Hill Creek watershed (fig. 5). The Pennsylvanian age rocks, from youngest to oldest, are the Casselman, Glenshaw, Allegheny, and Pottsville Formations. These rocks underlie most of the watershed. The Mississippian rocks, from youngest to oldest, are the Mauch Chunk Formation, the Loyalhanna Limestone, and the Burgoon Sandstone. These rocks are found in the higher elevations and underlie the western part of the watershed. The Loyalhanna Limestone is quarried for aggregate in places. The New Enterprise Stone and Lime Company operates two quarries in the Bakersville area.

From a spatial analysis of 2001 land cover (fig. 6), it was determined that the Laurel Hill Creek watershed is 63.4 percent forest, 27.2 percent agricultural, 4.9 percent residential, 3.0 percent wetlands and open water, and 1.6 percent commercial/industrial and mining (Scott Hoffman, U.S. Geological Survey, written commun., 2012). Together forested and agricultural areas account for more than 90 percent of the watershed. The upper one-third of the watershed is mainly agricultural. Agricultural land use predominantly coincides with the areas underlain by the Casselman and Glenshaw Formations.

Water Withdrawals and Use

Water is withdrawn from both surface and groundwater sources to supply multiple uses. In 2003, the average daily quantity of water withdrawn from the Laurel Hill Creek watershed was 2.27 million gallons per day (Mgal/d); in 2009, the average daily withdrawal was 2.22 Mgal/d. Fifty-two percent of the water withdrawn in 2003 was from surface-water sources, and 38 percent of the water was from groundwater sources. In 2003, 68 percent of the withdrawals was for public water supply, 14 percent was for commercial purposes, and 8 percent was for the mining industry (Pennsylvania Department of Environmental Protection, 2009).

The Laurel Hill Creek Lake (fig. 1) is owned by the Borough of Somerset (referred to as Somerset), a municipal water supplier whose service area is located outside and to the east of the Laurel Hill Creek watershed. Somerset also withdraws and exports water from three wells within the watershed. Somerset has a minimum pass-by requirement downstream from the reservoir of 1.37 Mgal/d. When the discharge drops below this minimum amount, Somerset enacts a drought plan in which the withdrawals from the reservoir cease and withdrawals from the water-supply wells increase. These changes in operating procedures result in an overall reduction in withdrawals, which is sufficient to meet water-supply demands over a prolonged period (Borough of Somerset, written commun., 2008). In 1994, 1995, 1998, 1999, 2001, and 2002, Somerset continued to withdraw water from the reservoir when discharge was less than the minimum pass-by

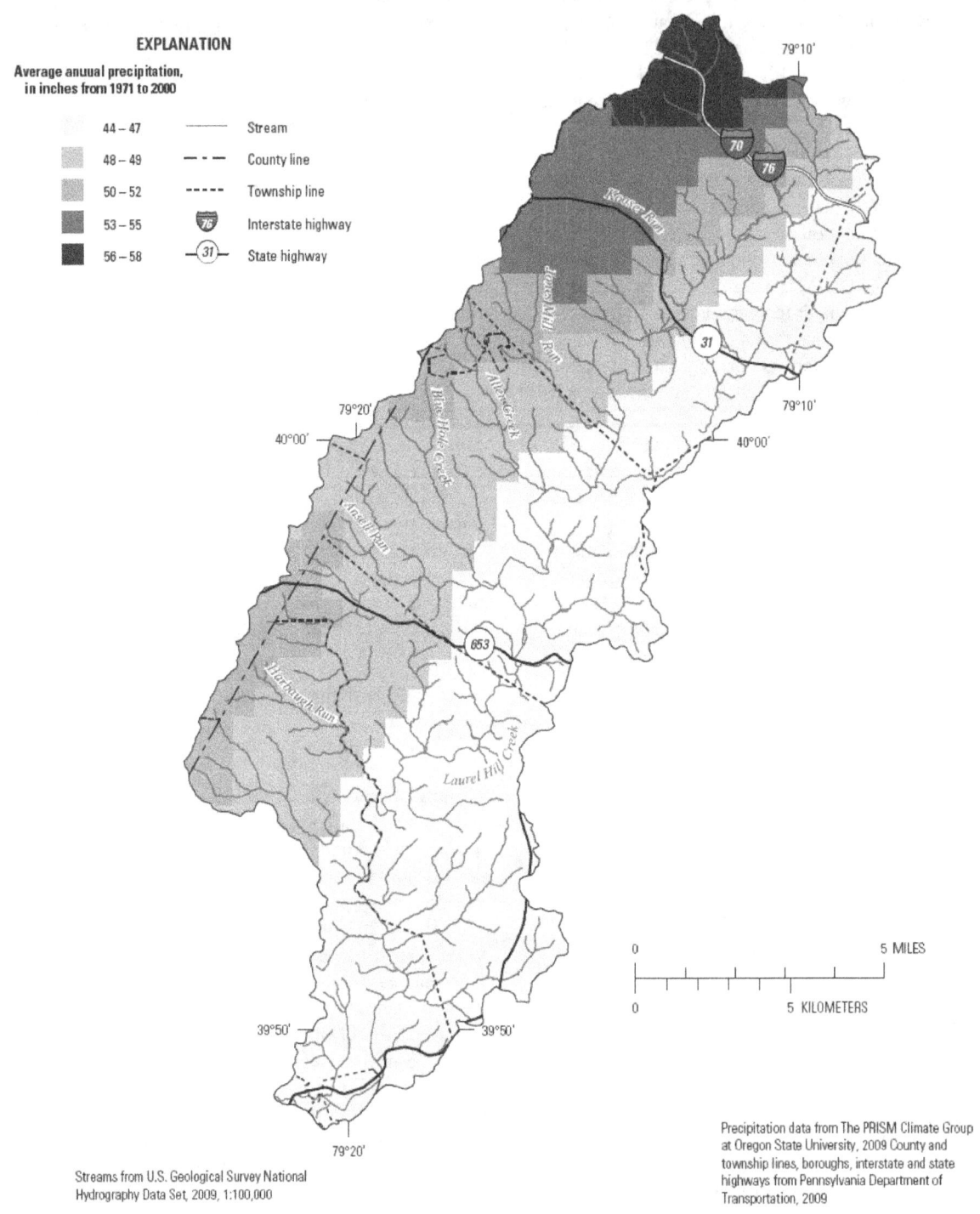

EXPLANATION

Average annual precipitation,
in inches from 1971 to 2000

44 – 47	Stream
48 – 49	County line
50 – 52	Township line
53 – 55	Interstate highway
56 – 58	State highway

Streams from U.S. Geological Survey National
Hydrography Data Set, 2009, 1:100,000

Precipitation data from The PRISM Climate Group
at Oregon State University, 2009 County and
township lines, boroughs, interstate and state
highways from Pennsylvania Department of
Transportation, 2009

Figure 3.   Average annual precipitation in the Laurel Hill Creek watershed, Somerset County, Pennsylvania, 1971–2000.

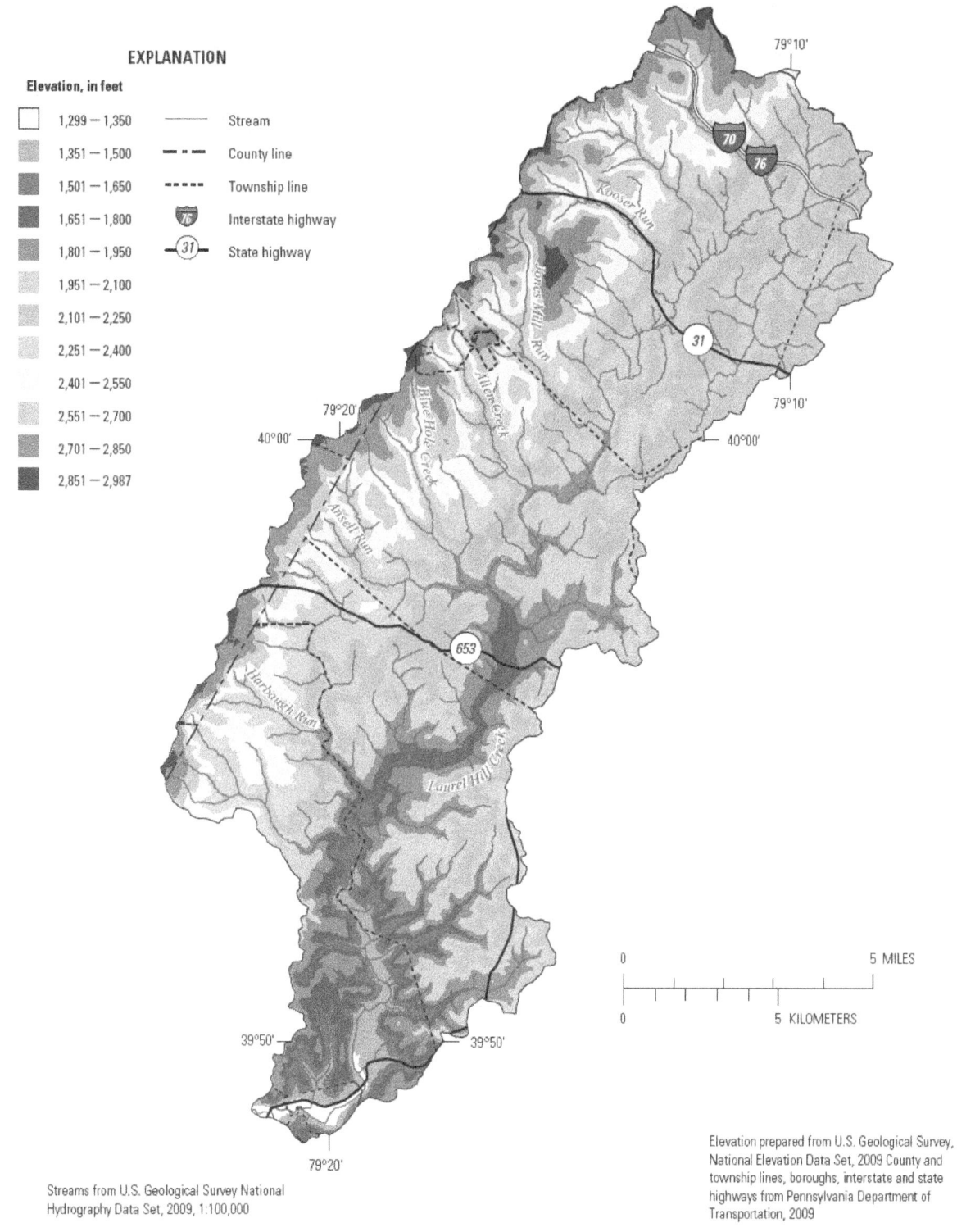

Elevation prepared from U.S. Geological Survey, National Elevation Data Set, 2009 County and township lines, boroughs, interstate and state highways from Pennsylvania Department of Transportation, 2009

Streams from U.S. Geological Survey National Hydrography Data Set, 2009, 1:100,000

Figure 4.   Topographic elevations in the Laurel Hill Creek watershed, Somerset County, Pennsylvania.

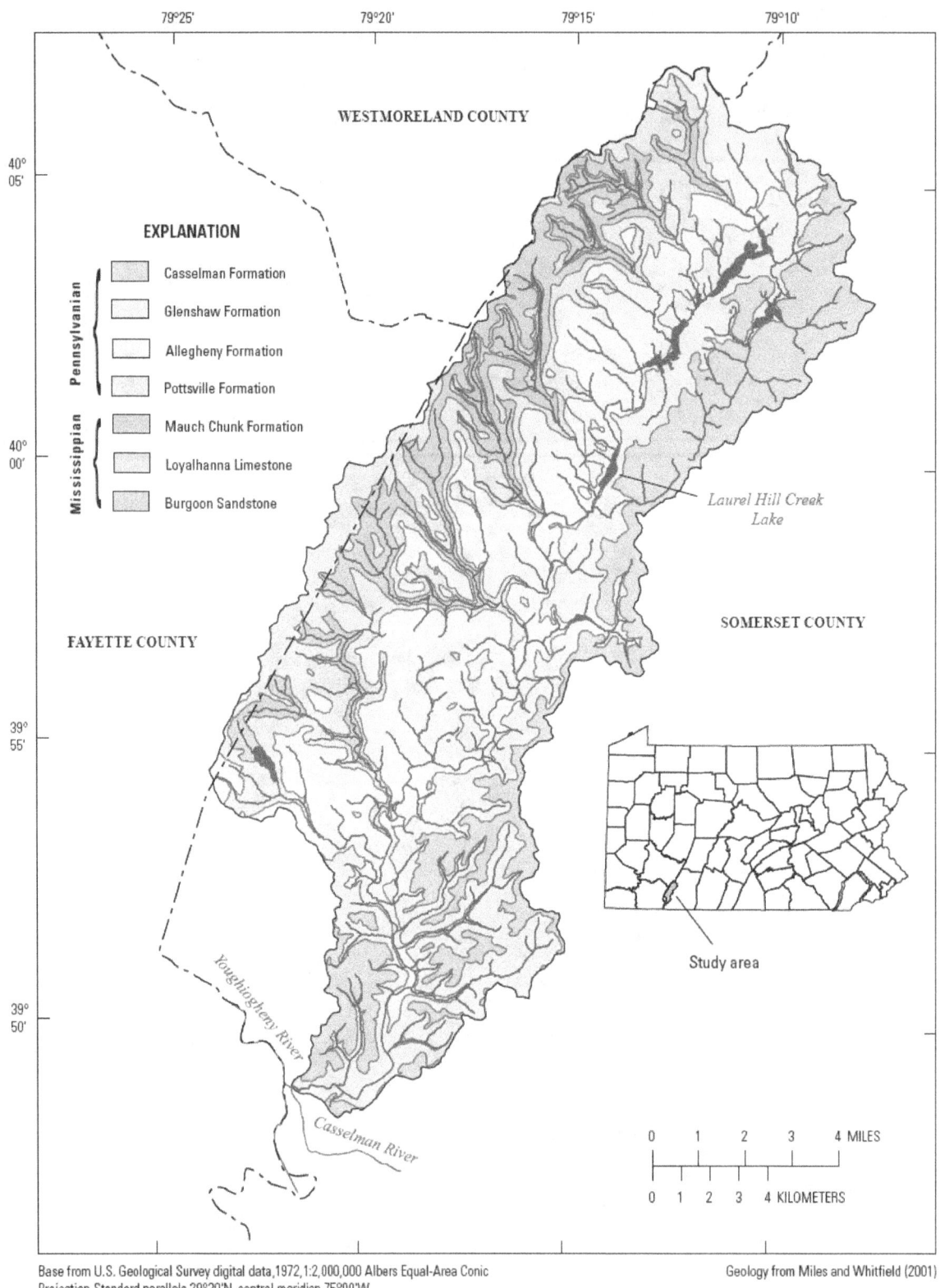

Figure 5. Geology of the Laurel Hill Creek watershed, Somerset County, Pennsylvania.

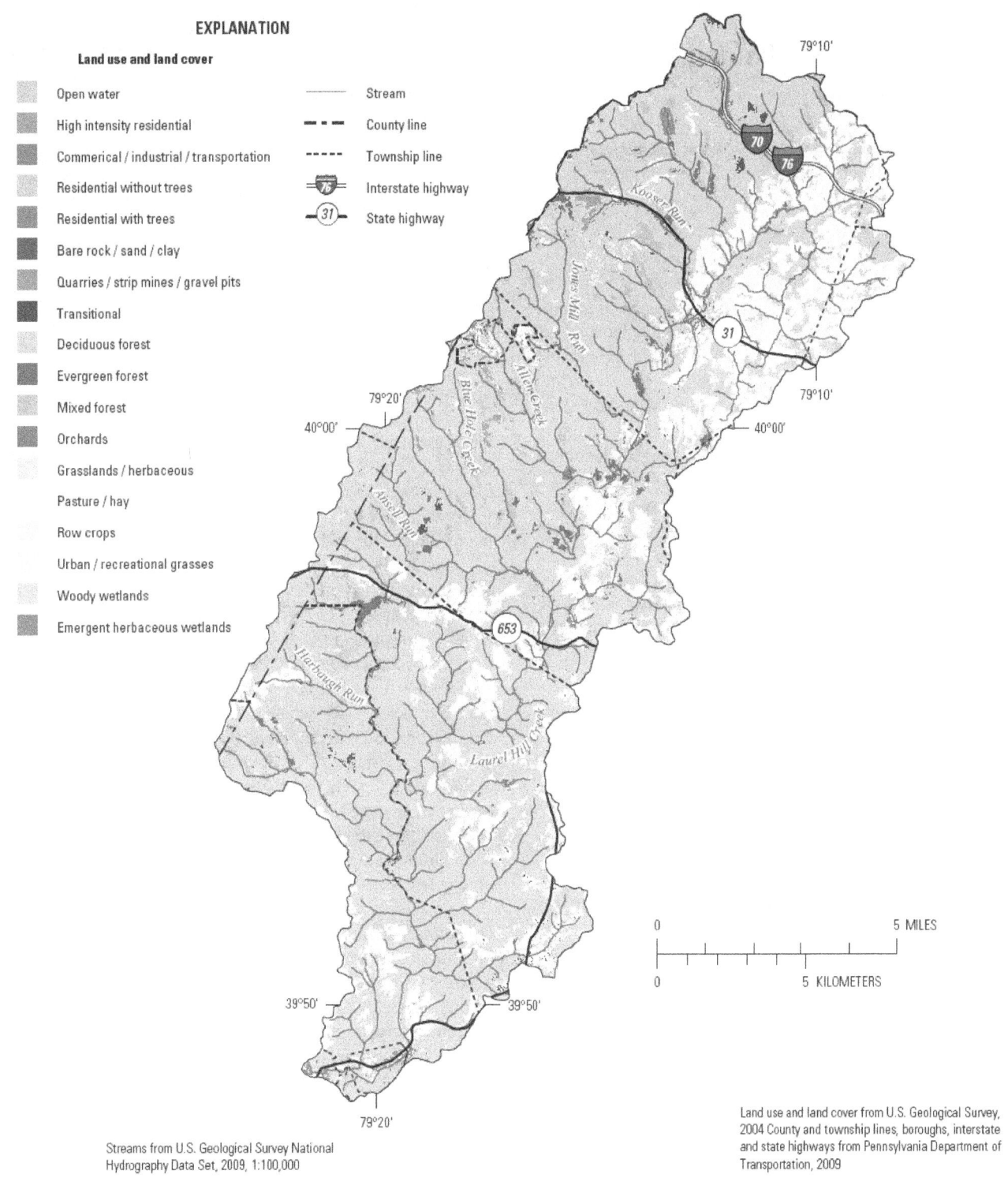

EXPLANATION

Land use and land cover

Open water

High intensity residential

Commerical / industrial / transportation

Residential without trees

Residential with trees

Bare rock / sand / clay

Quarries / strip mines / gravel pits

Transitional

Deciduous forest

Evergreen forest

Mixed forest

Orchards

Grasslands / herbaceous

Pasture / hay

Row crops

Urban / recreational grasses

Woody wetlands

Emergent herbaceous wetlands

Stream

County line

Township line

Interstate highway

State highway

Streams from U.S. Geological Survey National
Hydrography Data Set, 2009, 1:100,000

Land use and land cover from U.S. Geological Survey,
2004 County and township lines, boroughs, interstate
and state highways from Pennsylvania Department of
Transportation, 2009

Figure 6.   Land use in the Laurel Hill Creek watershed, Somerset County, Pennsylvania.

requirement of 1.37 Mgal/d (Pennsylvania Department of Environmental Protection, 2009).

Methods of Data Collection and Chemical Analysis

Methods for collection and analysis of discharge, turbidity, suspended-sediment, and nutrient data are described in the following sections.

Discharge

Discharge data were collected at two streamflow-gaging stations on Laurel Hill Creek ((03079600 and 03080000; fig. 1) at 15-minute intervals. Published data for these stations are available at *http://pa.water.usgs.gov/*.

The streamflow-gaging station on Laurel Hill Creek near Bakersville, Pa., (station 03079600) is in Laurel Hill State Park above Laurel Hill Creek Lake about 2.5 mi south-south-west of Bakersville. The drainage area is 38.2 mi². This station was installed in October 2009 to collect discharge, turbidity, nutrient, and suspended-sediment data. A turbidity probe and automatic water sampler were installed at this site.

The streamflow-gaging station on Laurel Hill Creek at Ursina, Pa., (station 03080000) is 500 ft downstream from the bridge on State Highway 281 at Ursina and 2.7 mi upstream from the mouth of the creek. The drainage area is 121 mi². Discharge data have been collected at this station since October 1918.

Turbidity

Turbidity data were collected at the Bakersville station (03079600). Data were collected at 15-minute intervals using an FTS DTS-12 turbidity sensor installed in a perforated 2-inch-diameter steel pipe. The sensor was positioned approximately 20 ft from the left bank, which is about mid-stream. The probe uses a 780 nanometer wavelength for a light source. The sensor measurement range is 0 to 1,600 Formazin Nephelometric Units (FNU). Accuracy is ±2 percent of the reading for 0 to 300 FNU and ±4 percent of the reading for 400 to 1,600 FNU. FTS, Inc., reports that the DTS-12 sensor exhibits less than 2-percent annual drift, eliminating the typical requirement for frequent turbidity probe calibration (FTS, Inc., 2011). The reporting precision of the turbidity probe in low turbidity water is ±0.3 FNU. The lowest reportable value, 0 FNU, does not necessarily indicate a condition of zero turbidity; the turbidly could be as great as 0.3 FNU. Published turbidity data are available at *http://pa.water.usgs.gov/*.

Suspended Sediment

Suspended-sediment samples were collected at the Bakersville, Ursina, and Trent (03079655) stations (fig. 1). Suspended-sediment data for the Bakersville station are listed in appendix 1, for the Ursina station in appendix 2, and for the Trent station in appendix 3.

Suspended-sediment samples were collected at the Bakersville station during high (storm) flows by an automatic sampler that was programmed to sample streamflow above a set gage height (trip stage). When flow was above the trip stage, one sample was collected every hour to fill 24 one-liter plastic bottles. Sampling ceased when the flow dropped below the trip stage or the 24 bottles were filled. Samples were collected during eight storms using the automatic sampler. The number of hourly samples for each storm ranged from 3 to 24.

A 1.2 liter (L) flow-weighted sample was processed for each storm. To generate a flow-weighted sample, the discharge at the time each sample was collected was determined from the continuous stage data using a rating curve. From the discharge data, the amount of the sample to be withdrawn from each bottle was calculated in order to have a total flow-weighted sample of 1.2 L for the storm. Samples withdrawn from each bottle were then composited. The discharge reported for each composited storm sample is the mean discharge (appendix 1). The composited sample was processed into other bottles for nutrient and sediment-concentration analyses. Because USGS protocols were not followed for sample compositing, the sediment concentration is reported as suspended solids in appendix 1.

Samples for suspended-sediment analysis were sent to the USGS Kentucky Water Science Center Sediment Laboratory. The concentration was determined by use of American Society for Testing and Materials (ASTM) method D3977-97 (American Society for Testing and Materials, 2002), and the entire sample was analyzed. For suspended-sediment samples collected for source tracking, the samples were centrifuged and passed through a 63-micron polyester sieve to remove sand and then sent to the appropriate laboratory for analysis.

During low streamflow (base flow), sediment samples from all three stations were collected in bottles placed in a DH-81 sediment sampler. Samples were collected by USGS personnel wading across the stream at a cross section near the streamflow-gaging station.

Some samples collected at the Ursina station were sent to the PaDEP laboratory in Harrisburg, Pa., for total suspended-solids (TSS) analysis. TSS is the concentration of total suspended material carried by a stream, as determined by an analysis of a representative subsample of a collected water sample. The analysis method requires the sample to be shaken, an aliquot taken, and only the aliquot analyzed, not the entire sample. A TSS analysis may underestimate the total suspended material if a significant fraction of the suspended load is sand-sized or larger. Because the analytical methods used to determine concentrations of suspended sediment and total suspended solids (TSS) differ, concentrations of suspended sediment tend to be higher, and measurements tend to be more accurate than those of TSS, particularly at higher flows (Kammerer and others, 1998). Since February 2001, USGS policy has mandated the use of suspended-sediment analysis rather than TSS analysis.

At the Ursina station, grab samples were collected at a cross section adjacent to the streamflow-gaging station by using same methods described above for the Bakersville station. Only one composite storm sample (September 30, 2010) was collected with an automated sampler. Storm samples collected for the Pennsylvania Water Quality Network (WQN) (Pennsylvania Department of Environmental Protection, 2006b) were discrete samples, and only one sample was collected for each storm. The WQN sampling frequency at the Ursina station is one sample per month. Five storm samples for the Ursina station were collected upstream from the streamflow-gaging station at the Route 281 bridge; the samples were depth- and width-integrated samples.

The Trent station does not have a streamflow-gaging station. Grab samples were collected in a manner similar to that for the other two stations. A discharge measurement was made at the time of sample collection.

Nutrients

Water samples were collected for nutrient analysis at all three sites. Samples were sent to either the USGS National Water Quality Laboratory in Denver, Colorado, or the PaDEP laboratory in Harrisburg, Pa. (see appendix 1, 2, and 3). Nutrients analyzed include dissolved ammonia, dissolved nitrite, dissolved nitrate plus nitrite, dissolved and total phosphorus, total dissolved nitrogen, and total nitrogen. For some samples from the Ursina station, nutrients analyzed also included total ammonia, dissolved nitrate, and total orthophosphate. Nutrient data are given for the Bakersville station in appendix 1, the Ursina station in appendix 2, and the Trent station in appendix 3.

Storm samples from the Bakersville station were collected using an automatic sampler, and a flow-weighted sample was processed for each storm as described in the preceding section. Samples were analyzed at the U.S. Geological Survey National Water Quality Laboratory in Denver, Colo. Monthly samples were collected during the 2011 water year at the Ursina station as part of the WQN. These samples were analyzed by the PaDEP laboratory in Harrisburg, Pa.

Sediment-Source Samples

The sediment-fingerprinting approach provides a direct method for determining the likely sources of watershed-derived fine-grained suspended sediment (Collins and others, 1997; Motha and others, 2003; Walling, 2005; Gellis and others, 2009). This approach entails the identification of specific sediment sources through the establishment of a minimal set of physical and (or) chemical properties, such as tracers that uniquely define each source in the watershed. Suspended sediment collected under different flow conditions exhibits a composite, or fingerprint, of properties that allows them to be traced back to their respective sources. Tracers that have been used successfully as fingerprints are minerals (Motha and others, 2003), radionuclides (Walling and Woodward, 1992; Collins and others, 1997; Nagle and others, 2007); trace elements (Devereux and others, 2010); magnetic properties (Slattery and others, 2000), and stable isotope ratios ($^{13}C/^{12}C$ and $^{15}N/^{14}N$) (Papanicolaou and others, 2003). Sources of sediment in a watershed include channel corridor (streambanks) and upland areas containing agriculture, urban construction, and forest. Sampling sediment at these sources and linking the fingerprints to sediment in transport using a statistical mixing model enables quantification of the sediment from each source. Sediment-source samples were collected upstream from the Bakersville station in upland source areas and streambanks. Sediment sources were identified as agriculture (cropland and pasture), forest, unpaved road, and streambank (fig. 7).

Site selection for sampling sediment in areas with forest and agriculture was based on (1) landowner permission and (2) the ability to obtain a spatially representative data set. Topographic maps (7.5 minute) available for the Laurel Hill Creek watershed display unpaved roads. After a reconnaissance of the watershed, samples from unpaved roads were collected to obtain a spatially representative data set. Unpaved roads were distinguished in the field as semi-paved sand or gravel, maintained sand and gravel, and non-maintained sand and gravel. Samples from unpaved roads were obtained by sweeping the surface sediment with a small broom into a plastic pan. Approximately 150 ft of road was sampled at each site.

Site selection for sampling streambanks was based on a spatial analysis of streams in the Laurel Hill Creek watershed. The streams were classified by Strahler order into first, second, third, and fourth order (Strahler, 1952), and the lengths of streams in each order were summed (table 1). The number of samples collected for each stream order was based on the length of streams in each order, as well as field observations of eroding streambanks. Spatially distributed sampling sites were then selected. To obtain a representative sample, the streambanks were sampled from the bottom to the top of the bank face. Samples were collected at three to five transects spaced 30 ft apart and composited into one sample. If streambanks were exposed on both sides of the channel, samples were collected from both sides of the stream and composited into one sample.

Soil samples from agricultural and forested areas were collected from the soil surface with a plastic hand shovel. To account for variability in the tracer properties at agricultural and forested sites, sediment was collected across transects and composited into one sample. Transects were typically 350 by 100 ft.

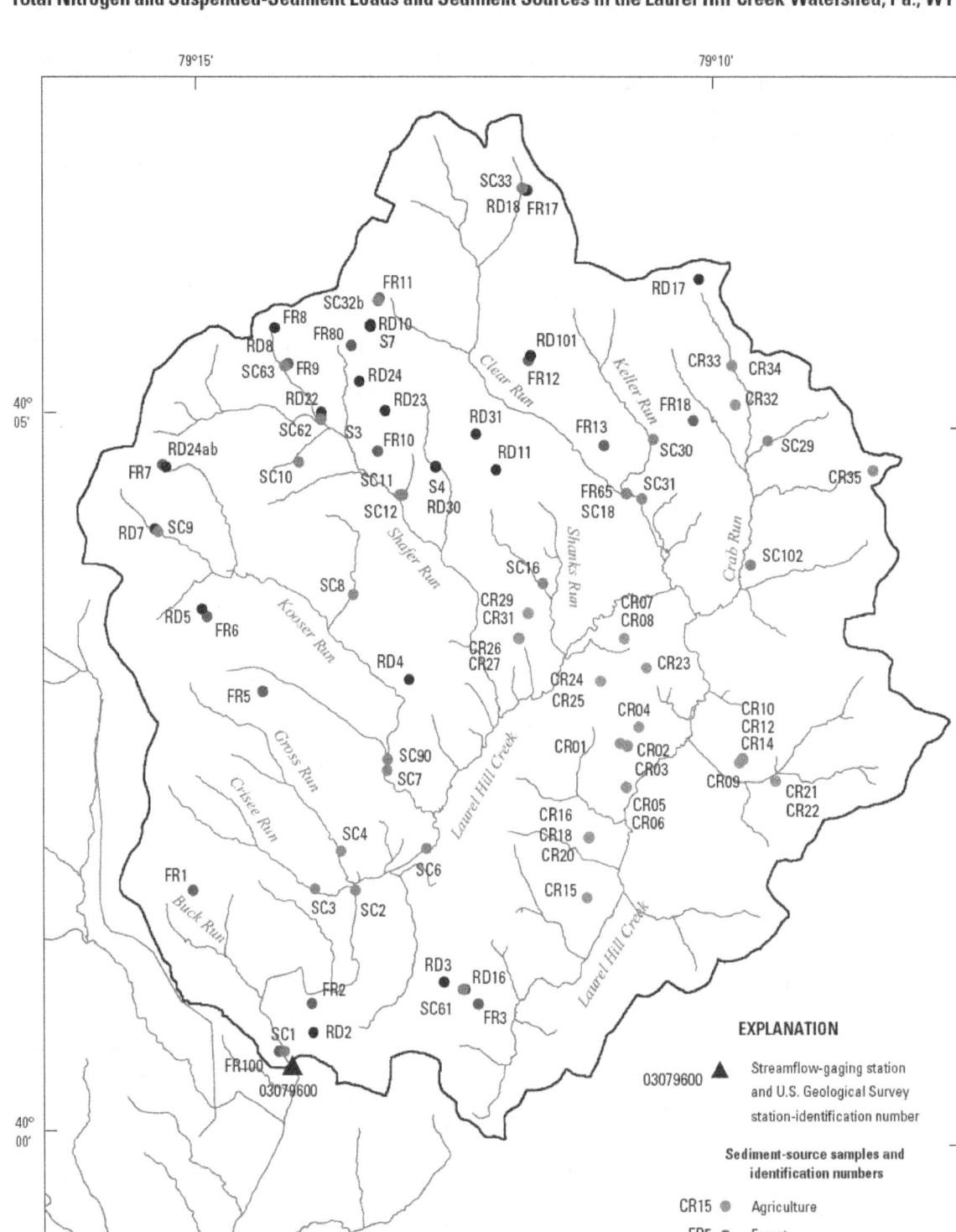

Figure 7. Location of sediment-source sample sites in the Laurel Hill Creek watershed, Somerset County, Pennsylvania.

Table 1.   Summed lengths of stream reaches in the Laurel Hill Creek watershed, Somerset County, Pennsylvania, by stream order.

Stream order	Sum of lengths (feet)	Percentage of total stream length
1	239,390	54
2	121,730	27
3	36,350	8
4	49,400	11
Total	446,870	100

Laboratory Analyses for Sediment Fingerprinting

Samples collected from agricultural lands, forests, unpaved roads, and streambanks were taken to the laboratory, dried at 60 degrees Celsius (°C) (Gellis and others, 2009), disaggregated using a pestle and mortar, and wet-sieved through a 63-micron polyester sieve to remove the sand. Sample weights before and after sieving were recorded to determine the percentage of sand in the samples.

The silt and clay fractions (less than 63 microns) of suspended sediment, upland soil, and channel corridor samples were sent to a USGS research laboratory in Denver, Colo., for elemental analyses (table 2) and to the University of California at Davis Stable Isotope Laboratory for stable isotope analyses (table 3). At the USGS laboratory, the samples were analyzed for 35 elements (table 2). All samples were analyzed using inductively coupled plasma combined with mass spectrometry (ICP-MS) after multi-acid decomposition (a mixture of hydrochloric, nitric, perchloric, and hydrofluoric acids). Specific details regarding this method can be found in Taggart (2002).

Samples were analyzed for carbon and nitrogen stable isotopes ($^{13}C/^{12}C$ and $^{15}N/^{14}N$), total carbon (C), and total nitrogen (N) at the University of California laboratory (table 3) using an Elementar Vario EL Cube (ELEMENTAR Analysensysteme, GmbH, Hanau, Germany) elemental analyzer interfaced to a Sercon 20-20 isotope ratio mass spectrometer (Sercon Ltd., Cheshire, UK). Samples were combusted at 1,000°C in a reactor packed with cerium dioxide, copper oxide, and lead chromate. Following combustion, oxides were removed in a reduction reactor (reduced copper at 650°C). Water was removed with magnesium perchlorate. Carbon dioxide was removed from the carrier stream by an adsorption trap allowing nitrogen to be analyzed. Following the completion of the nitrogen analysis, the adsorption trap was heated releasing the trapped carbon dioxide for analysis (Brenna and others, 1997).

Carbon and nitrogen isotope values ($\delta^{13}C$ and $\delta^{15}N$) are reported in per mil (‰) notation with respect to Vienna Pee Dee Belemnite (VPDB) and atmospheric N_2 (AIR),

respectively. During analysis, samples were interspersed with laboratory standards, which were previously calibrated against International Atomic Energy Agency (IAEA) Standard Reference Materials including USGS-40, USGS-41, and IAEA-600. Provisional isotope values were normalized using USGS-41 ($d^{13}C_{VPDB}$ = 37.63‰ and $d^{15}N_{AIR}$ = 47.6‰) and an internal nylon standard ($d^{13}C_{VPDB}$ = -27.81‰ and $d^{15}N_{AIR}$ = -9.8‰). The long-term precision for a laboratory check standard is 0.2 ‰ for $\delta^{13}C$ and 0.3 ‰ for $\delta^{15}N$. The samples analyzed for carbon isotopes were acid fumed according to the procedures of Harris and others (2001).

Quality Control

Two sources of errors were determined for the fingerprint sample set. One source of error was from field sampling. Field sampling error was determined by collecting replicates of the source samples at selected locations. The replicate samples were collected at the same location with the same sampling methodology and sent to the appropriate laboratories for analysis. The difference in analytical results from two replicate samples shows possible errors in sampling. Another source of error is analytical error. Analytical errors were discerned by taking a split of the field sample. The difference in analytical results from the split samples shows possible errors in analytical results. Because of constraints in mass requirements, it was not possible to obtain replicates or splits of the suspended-sediment samples.

Replicate and split errors were determined as percent difference in the tracer values using the following equation:

$$\text{Error (percent)} = |(A - B)/((A + B)/2)| * 100 \ , \qquad (1)$$

where

A and B are the tracer values for a replicate or split sample.

When sample splits or replicates were analyzed, the tracer values used in the statistical analysis of sediment sources are the average of the splits of the sample and the replicate value for that sample.

The results of quality control for the sampling methodology (replicate samples) and laboratory analysis (split samples) are provided in appendix 4. A total of 9 split and 9 replicate samples were collected for the source groups (appendix 4). With the exception of forest sample FR80, which had an error of 200 percent, errors ranged from 0 to 30 percent for the split samples with 67 percent of the tracers having an error less than 10 percent. The largest errors in the split-sample group were for total carbon and total nitrogen (appendix 4). Errors for the replicate samples ranged from 0 to 65 percent, with 83 percent of the tracers having an error less than 10 percent (appendix 4).

Table 2.   Elements used for fingerprint analysis, denoted by name and symbol, analyzed at the U.S. Geological Survey Geology Discipline research laboratory in Denver, Colorado.

Aluminum (Al)	Cesium (Cs)	Lithium (Li)	Rubidum (Rb)	Vanadium (V)
Antimony (Sb)	Chromium (Cr)	Magnesium (Mg)	Scandium (Sc)	Yttrium (Y)
Arsenic (As)	Cobalt (Co)	Manganese (Mn)	Sodium (Na)	Zinc (Zn)
Barium (Ba)	Copper (Cu)	Molybdenum (Mo)	Strontium (Sr)	
Beryllium (Be)	Gallium (Ga)	Nickel (Ni)	Thalium (Tl)	
Bismuth (Bi)	Iron (Fe)	Niobium (Nb)	Thorium (Th)	
Cadmium (Cd)	Lanthanum (La)	Phosphorus (P)	Titanium (Ti)	
Cerium (Ce)	Lead (Pb)	Potassium (K)	Uranium (U)	

Table 3.   Tracer constituents analyzed by the University of California at Davis Stable Isotope Laboratory, Davis, California.

Stable isotope delta carbon-13 ($\delta^{13}C_{VPDB}$ (‰))

Percentage of dried soil sample comprised of total Carbon (C)

Stable isotope delta nitrogen-15 ($\delta^{15}N_{AIR}$ (‰))

Percentage of dried soil sample comprised of total Nitrogen (N)

Nutrient Concentrations and Loads

Water samples for nutrient analysis were collected at the Bakersville and Ursina stations from July 2009 through September 2011 (appendix 1 and 2). Nutrients analyzed included dissolved and total ammonia, dissolved and total nitrite, dissolved nitrate plus nitrite, dissolved nitrate, dissolved and total phosphorus, total orthophosphate, and dissolved and total nitrogen. Summary statistics for nutrients in water samples are presented in table 4.

Concentrations of nutrients in base-flow samples collected at the Bakersville station generally were low. Concentrations of dissolved ammonia ranged from less than 0.01 to 0.128 mg/L; the median concentration was 0.013 mg/L. All concentrations of dissolved phosphorus were less than the detection limit. Concentrations of total phosphorus ranged from less than 0.01 to 0.01 mg/L; only 2 of 10 samples had concentrations greater than the detection limit. Concentrations of dissolved nitrite ranged from an estimated 0.001 to 0.011 mg/L. Concentrations of nitrate plus nitrite ranged from 0.5 to 0.99 mg/L; the median concentration was 0.68 mg/L (table 4).

Concentrations of nutrients in stormflow composite samples collected at the Bakersville station generally were slightly higher than concentrations in baseflow. Concentrations of dissolved ammonia ranged from less than 0.01 to 0.046 mg/L; the median concentration was 0.024 mg/L. All concentrations of dissolved phosphorus except one (0.01 mg/L) were below the detection limit. Concentrations of total phosphorus ranged from less than 0.01 to 0.11 mg/L; the median concentration was 0.06 mg/L. Concentrations of dissolved nitrite ranged from 0.001 to 0.004 mg/L. Concentrations of nitrate plus nitrite ranged from 0.37 to 0.94 mg/L; the median concentration was 0.7 mg/L (table 4).

At the Bakersville station, concentrations of dissolved total nitrogen in base flow ranged from 0.61 to 1.3 mg/L in base-flow samples, 1.1 to 1.3 mg/L in stormflow grab samples, and 0.52 to 1.1 mg/L in stormflow composite samples. Median concentrations for base-flow and storm composite samples were similar at 0.88 and 0.91 mg/L, respectively. The median concentration for storm grab samples was 1.2 mg/L. Concentrations of total nitrogen ranged from 0.63 to 1.3 mg/L in base-flow samples, 1.0 to 1.4 mg/L in storm grab samples, and 0.57 to 1.5 mg/L in storm composite samples. Median concentrations were 0.88 mg/L in base-flow samples, 1.2 mg/L in storm grab samples, and 1.2 mg/L in storm composite samples (table 4).

Concentrations of nutrients in base-flow samples collected at the Ursina station generally were low. Concentrations of dissolved ammonia ranged from less than 0.02 to 0.069 mg/L; the median concentration was below the detection limit. Only one of nine samples had a total ammonia concentration (0.04 mg/L) greater than the detection limit. All concentrations of dissolved phosphorus were less than the detection limit. Concentrations of total phosphorus ranged from less than 0.02 to 0.02 mg/L; only 4 of 21 samples had concentrations greater than the detection limit. Concentrations of orthophosphate ranged from less than 0.01 to 0.02 mg/L; the median concentration was less than the detection limit. All concentrations of total nitrite were less than the detection limit

Table 4. Summary statistics for nutrients in water samples collected from three sites in Laurel Hill Creek watershed, Somerset County, Pennsylvania, 2009–11.

[all concentrations given in milligrams per liter; N, nitrogen; P, phosphorus; --, unable to compute statistic; NA, not analyzed; <, less than; E estimated concentration]

Station and sample type	Statistic	Ammonia, dissolved, as N	Ammonia, total, as N	Nitrate plus nitrite, dissolved, as N	Nitrite, dissolved, as N	Nitrite, total, as N	Nitrate, dissolved, as N	Phosphorus, dissolved	Phosphorus, total	Ortho-phosphate, total, as P	Total nitrogen, dissolved	Total nitrogen, total
Bakersville base-flow samples	Number of samples	10	NA	10	10	NA	NA	8	10	NA	10	10
	Range	<0.01–0.128	NA	0.5–0.99	0.001 E–0.011	NA	NA	<0.02	<0.01–0.01	NA	0.61–1.3	0.63–1.3
	Mean	--	NA	0.68	0.004	NA	NA	--	--	NA	0.89	0.88
	Median	0.013	NA	0.68	0.004	NA	NA	<0.02	<0.02	NA	0.88	0.88
Bakersville storm grab samples	Number of samples	3	NA	3	3	NA	NA	3	3	NA	3	3
	Range	0.014–0.051	NA	0.95–1.16	0.002 E–0.004	NA	NA	<0.02	<0.02–0.05	NA	1.1–1.3	1.0–1.4
	Mean	0.027	NA	1.03	0.00	NA	NA	--	--	NA	1.2	1.2
	Median	0.015	NA	0.97	0.00	NA	NA	<0.02	0.02	NA	1.2	1.2
Bakersville storm composite samples	Number of samples	8	NA	8	8	NA	NA	8	8	NA	8	8
	Range	<0.01–0.046	NA	0.37–0.94	0.001–0.004	NA	NA	<0.01–0.02	0.01–0.11	NA	0.52–1.1	0.57–1.5
	Mean	--	NA	0.64	0.00	NA	NA	--	0.06	NA	0.84	1.0
	Median	0.024	NA	0.7	0.003	NA	NA	<0.015	0.06	NA	0.91	1.2
Ursina base-flow samples	Number of samples	12	9	12	12	9	9	11	21	9	21	12
	Range	<0.02–0.069	<0.02–0.04	0.24–0.77	<0.002–0.007	<0.04	0.06–0.97	<0.02	<0.02–0.02 E	<0.01–0.02	0.12–1.1	0.25–0.92
	Mean	--	--	0.38	--	--	0.53	--	--	--	0.55	0.55
	Median	<0.02	<0.02	0.305	0.001	<0.04	0.56	<0.02	<0.01	<0.01	0.58	0.57
Ursina storm grab samples	Number of samples	2	3	2	2	3	3	2	5	3	5	2
	Range	<0.02–0.21	<0.02–0.04	0.33–0.87	0.001 E–0.002 E	<0.04	0.3–0.61	<0.02	0.01–0.24	<0.01–0.01	0.52–0.97	0.97–1.0
	Mean	--	--	--	--	--	--	--	0.063	--	0.69	--
	Median	--	0.04	--	--	--	--	--	0.02	<0.01	0.7	--

of 0.04 mg/L. Concentrations of dissolved nitrite ranged from less than 0.002 to 0.007 mg/L; the median concentration was 0.001 mg/L. Concentrations of nitrate plus nitrite ranged from 0.24 to 0.77 mg/L; the median concentration was 0.31 mg/L (table 4).

At the Ursina station, concentrations of dissolved total nitrogen ranged from 0.12 to 1.1 mg/L in 21 base-flow samples and from 0.52 to 0.97 mg/L in 5 stormflow samples. Median concentrations for base-flow and stormflow samples were similar at 0.58 and 0.7 mg/L, respectively. Concentrations of total nitrogen ranged from 0.25 to 0.92 mg/L in 12 base-flow samples and were 0.97 and 1.0 mg/L in 2 stormflow samples. The median concentration for base-flow samples was 0.57 mg/L (table 4).

The term "load" represents the mass (commonly expressed in tons or pounds) of a constituent transported past a sampling station during a specified period of time. Loads can be computed for various time increments, such as instantaneous, daily, monthly, seasonal, and annual, or for storms. Instantaneous loads represent the mass transported at the specific sampling time, whereas daily, monthly, seasonal, annual, and storm loads represent the cumulative mass transported over a prolonged period.

Monthly total nitrogen loads were estimated for the Bakersville station by summing the daily estimated total nitrogen load. A regression model was developed to relate the total nitrogen measured in composite stormflow samples to the mean discharge during collection of the sample (fig. 8). The coefficient of determination was 0.93. The daily total nitrogen load was estimated for stormflows using

$$\log(N) = 0.00673 \, (0.2727 \log(Q) - 0.611) \; , \qquad (2)$$

where

  N is daily total nitrogen load, in pounds per day;

  Q is mean daily discharge, in cubic feet per second; and 0.00673 is a conversion factor.

A Duan bias correction factor (Duan, 1983) of 1.003 was applied to correct for negative bias during retransformation of the response variable.

A relation between total nitrogen concentration and mean discharge was not apparent for base-flow samples collected at the Bakersville station (fig. 9). Therefore, the mean total nitrogen concentration of the base-flow samples, 0.88 mg/L (table 4), was used to estimate total nitrogen loads during base flow. The daily total nitrogen load was estimated using

$$N = 0.00673 \, (0.88 \, Q) \; , \qquad (3)$$

where

  N is daily total nitrogen load, in pounds per day;

  Q is mean daily discharge, in cubic feet per second; and 0.00673 is a conversion factor.

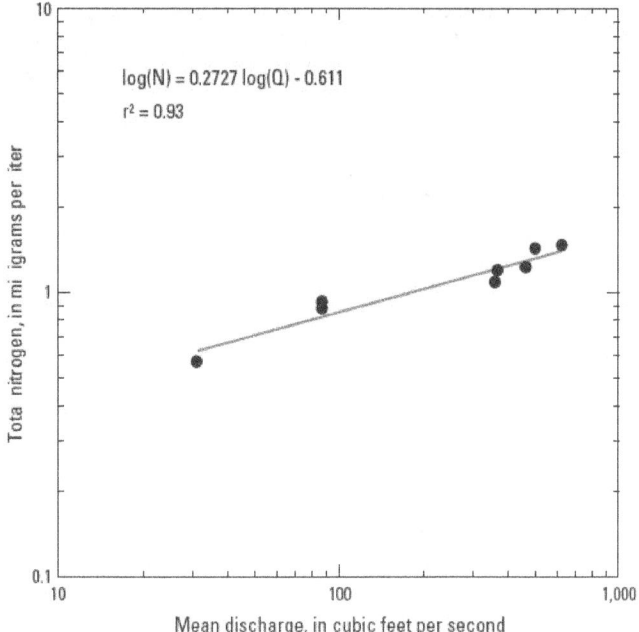

Figure 8. Regression relation of total nitrogen concentration to mean discharge for composite stormflow samples collected at station 03079600, Laurel Hill Creek near Bakersville, Pennsylvania.

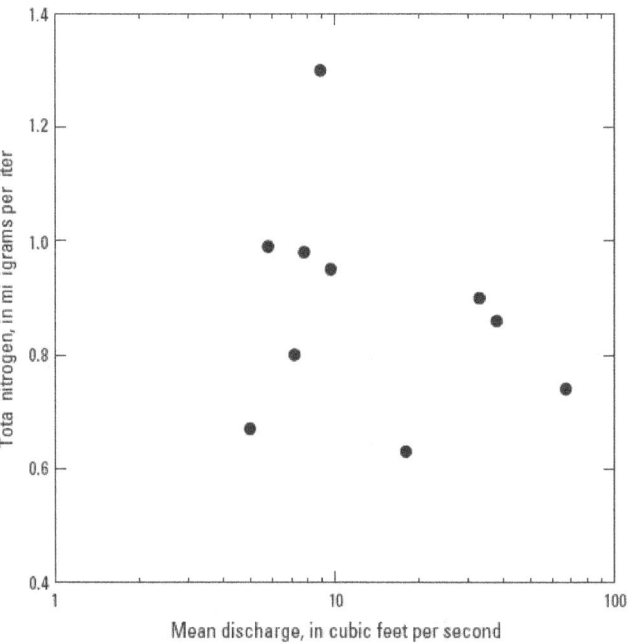

Figure 9. Relation of total nitrogen concentration to discharge for base-flow samples collected at station 03079600, Laurel Hill Creek near Bakersville, Pennsylvania.

Estimated monthly total nitrogen loads for the Bakersville station are given in table 5. The total nitrogen load was 262 pounds (lb) for 11 months of the 2010 water year and 266 lb for the 2011 water year. Mean monthly total nitrogen loads were similar for the 2010 and 2011 water years. The mean monthly total nitrogen load was 23.9 pounds per month (lb/mo) for the 2010 water year and 22.2 lb/mo for the 2011 water year. Most of the total nitrogen load was from stormflows. The mean monthly stormflow load was 18.3 lb/mo for the 2010 water year and 17.9 lb/mo for the 2011 water year. The stormflow load made up 76.6 percent of the total load for the 2010 water year and 80.6 percent of the total load for the 2011 water year (table 5).

The estimated monthly total nitrogen loads at the Bakersville station were higher during the winter and spring (December through May) than during the summer (June through August) (fig. 10). The estimated monthly total nitrogen load is related to the mean monthly discharge (fig. 11). The monthly total nitrogen load can be estimated from the mean monthly discharge using

$$N_m = 0.1226 \, Q^{1.1079} \, , \qquad (4)$$

where

  N is monthly total nitrogen load, in pounds per month; and

  Q is mean monthly discharge, in cubic feet per second.

Water samples were collected at both the Trent (station 03079655; appendix 3) and Bakersville (station 03079600; appendix 1) stations on 3 days (July 28, 2009; September 3, 2009, and June 7, 2010). The Bakersville station is located upstream from Laurel Hill Creek Lake, and the Trent station is located downstream from Laurel Hill Creek Lake. A comparison of the analytical results showed that the discharge and water temperature were higher downstream from Laurel Hill Creek Lake than upstream. The dissolved oxygen concentration, pH, specific conductance, dissolved nitrate plus nitrite, and total and dissolved nitrogen were lower downstream from Laurel Hill Creek Lake.

Suspended-Sediment Concentrations and Loads

Turbidity is a principal physical characteristic of water and is an expression of the optical property that causes light to be scattered and absorbed by particles and molecules rather than transmitted in straight lines through a water sample. It is caused by suspended matter or impurities that interfere with the clarity of the water. These impurities may include clay, silt, finely divided inorganic and organic matter, soluble colored organic compounds, and microscopic organisms (U.S. Environmental Protection Agency, 1999, p. 7–1). Suspended

Table 5. Estimated monthly total nitrogen loads for station 03079600, Laurel Hill Creek near Bakersville, Pennsylvania, November 2009 to September 2011.

[lb/mo; pounds per month]

Month and year	Base-flow load (lb/mo)	Stormflow load (lb/mo)	Total load (lb/mo)
November 2009	6.3	1.5	7.8
December 2009	6.8	21.9	28.7
January 2010	13.5	40.4	53.8
February 2010	9.5	0.7	10.2
March 2010	3.4	101	104
April 2010	9.6	5.7	15.3
May 2010	3.6	26.9	30.6
June 2010	4.8	2.4	7.2
July 2010	1.7	0.3	2.0
August 2010	1.3	0.0	1.3
September 2010	0.8	0.3	1.1
Total (11 months)	61.3	201	262
Mean	5.6	18.3	23.9
Percentage	23.4	76.6	100.0
October 2010	1.7	0.5	2.2
November 2010	2.9	11.2	14.1
December 2010	8.8	16.2	25.0
January 2011	2.6	2.7	5.3
February 2011	6.0	44.5	50.4
March 2011	8.4	44.5	53.0
April 2011	0.8	41.1	41.9
May 2011	7.7	16.8	24.5
June 2011	4.0	1.6	5.6
July 2011	1.9	0.7	2.6
August 2011	2.4	1.7	4.0
September 2011	4.4	33.1	37.5
Total (12 months)	51.59	215	266
Mean	4.3	17.9	22.2
Percentage	19.4	80.6	100

solids affect the scattering of light, and turbidity increases as the scattering of light increases.

Turbidity or turbidity and discharge are often better indicators of suspended-sediment concentration (SSC) than the traditional sediment-transport curve method for some streams. Turbidity can serve as a surrogate for SSC. Lee and others (2008) compared annual suspended-sediment loads (SSL) computed using traditional sediment-transport curve methods and a turbidity-SSC model at stations near John Redmond Reservoir in Kansas. Lee and others (2008) found that the

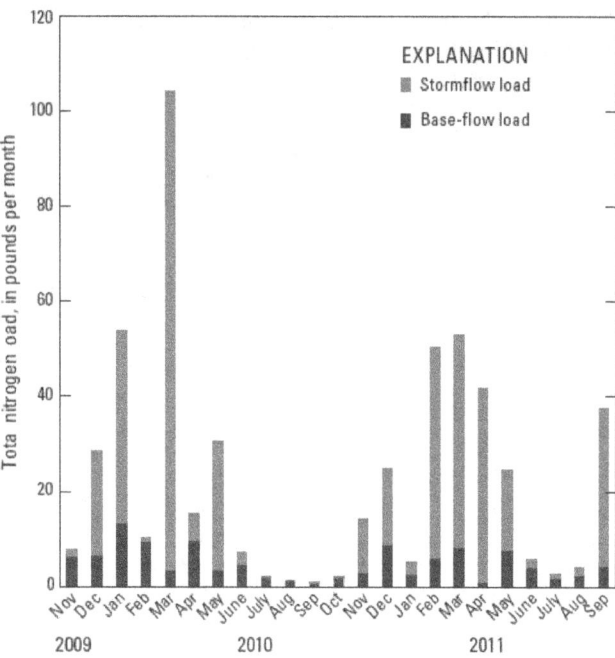

Figure 10. Estimated monthly total nitrogen loads for station 03079600, Laurel Hill Creek near Bakersville, Pennsylvania, November 2009 to September 2011.

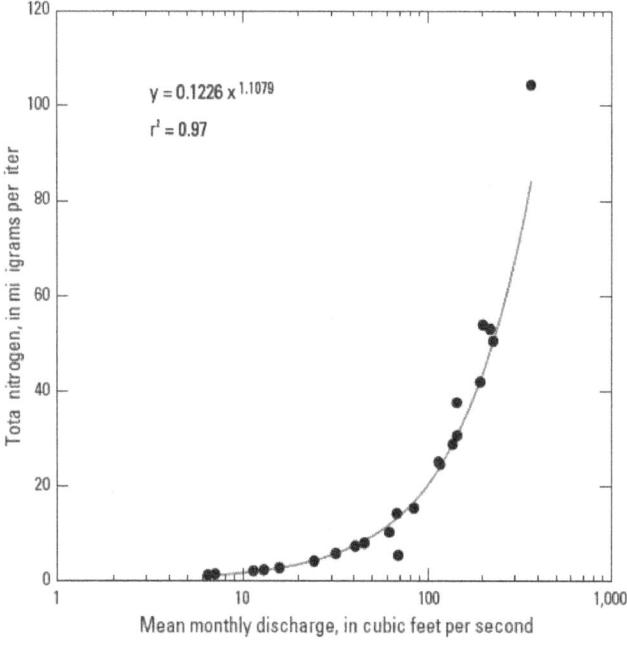

Figure 11. Relation between estimated monthly total nitrogen load and mean monthly discharge at station 03079600, Laurel Hill Creek near Bakersville, Pennsylvania, November 2009 to September 2011.

SSL calculated from the turbidity-SSC model had an error of 1.1 to 3.2 percent, whereas the SSL calculated from sediment-transport curves had an error of 16 to 20 percent.

Regression Models

Regression models were developed for SSC, turbidity, and discharge as a means of estimating time-series SSC using turbidity data. Available data from the 2010 and 2011 water years were used. The estimated SSC was used to compute daily, storm, and annual SSL for water years 2010 and 2011. Regression models were developed to determine the relations of SSC to turbidity (T) and discharge (Q) using SSC data collected by the automated sampler and turbidity and discharge data collected at the Bakersville station. The methodology is described by Rasmussen and others (2009), Helsel and Hirsch (2002), and Bragg and others (2007). Regression equations were developed using the model building approaches explained by Helsel and Hirsch (2002).

Six regression models were initially developed: (1) SSC in relation to T, (2) SSC to Q, (3) SSC to T and Q, (4) $log_{10}SSC$ to $log_{10}T$, (5) $log_{10}SSC$ to $log_{10}Q$, and (6) $log_{10}SSC$ to $log_{10}T$ and $log_{10}Q$. Statistics, including coefficient of determination (R^2), adjusted R^2, standard error, prediction error sum of squares (PRESS), Mallow's Cp, variance inflation factor, probability plot correlation coefficient, and model standard percentage error, were used to evaluate the models (table 6). The statistics are appropriate only for comparing models with the same response variable units. The coefficient of determination is the fraction of the variance explained by the regression model. The adjusted coefficient of determination is adjusted for the number of degrees of freedom to allow comparison of regression models with differing numbers of explanatory variables. The standard error is an estimate of the variation from the average. The prediction error sum of squares (PRESS) estimates error by using n-1 observations in the regression model to estimate the value left out; a lower PRESS value indicates less model error. Mallow's Cp is used to assess the fit of a regression model by minimizing bias and standard error; the best model is the one with the lowest Cp. The variance inflation factor (VIF) estimates how much the variance of an estimated regression coefficient in a multiple linear regression model is increased because of collinearity. The probability plot correlation coefficient (PPCC) is a test for normal distribution, which will have a correlation coefficient close to 1.0. As data depart from normality, the coefficient drops below 1.0. The model standard percentage error (MSPE) is the root mean squared error expressed as a percentage. It is a measure of the variance between observed values and values computed by the regression model. A log transformation model using turbidity and discharge as explanatory variables provided the best results (model number 6 in table 6). The regression model selected was

Table 6. Statistics for suspended-sediment regression models developed for station 03079600, Laurel Hill Creek near Bakersville, Pennsylvania, with suspended sediment concentration as the response variable.

[R^2, coefficient of determination; PPCC, probability plot correlation coefficient; PRESS, prediction error sum of squares; MSPE, model standard percentage error; VIF, variance inflation factor; -- statistic not applicable; n, number of observations; T, turbidity; Q, discharge]

Model number	Explanatory variable	R^2	Adjusted R^2	Standard error	PPCC	PRESS	Mallow's Cp	MSPE	VIF
	n = 37								
1	T	0.847	0.843	26.17	0.97	27,674	2.00	37.87	--
2	Q	0.637	0.627	40.35	0.99	62,546	50.20	58.38	--
3	T, Q	0.852	0.843	26.17	0.97	27,821	3	--	3.16
4	$\log_{10}T$	0.854	0.850	0.28	0.96	3.17	7.06	69.54	--
5	$\log_{10}Q$	0.761	0.754	0.36	0.97	1.30	4.94	32.68	--
6	$\log_{10}T$ and $\log_{10}Q$	0.876	0.869	0.26	0.96	2.94	3	--	3.90

$$\log(SSC) = 0.82 \log(T) + 0.362 \log(Q) - 0.459 \ , \quad (5)$$

where

- SSC is instantaneous suspended-sediment concentration, in milligrams per liter;
- T is turbidity, in Formazine Nephelometric Units; and
- Q is instantaneous discharge, in cubic feet per second.

Estimation of Suspended-Sediment Loads and Yields

Annual loads generally are more informative than instantaneous loads measured at the time of sampling because they represent the cumulative transport of sediment during a year and, thereby, incorporate the potentially large range of daily variations. Differences in annual loads transported past a sampling station can result from differences in annual flow volumes, physical watershed characteristics, current and historical land-use activities, and local conditions that affect sediment supply or susceptibility to erosion.

Instantaneous (15-minute) suspended-sediment loads (SSL) were computed from time-series turbidity and discharge data for the 2010 and 2011 water years (figs. 12 and 13, respectively) as follows. Equation 5 was used to compute instantaneous SSC from turbidity and discharge data collected every 15 minutes. SSC values computed from regression estimates were multiplied by the corresponding discharge values and a unit conversion to compute estimates of instantaneous SSL in short tons (2,000 lb). Instantaneous SSL was calculated using the following equation:

$$SSL_i = SSC_i \times Q_i \times C \ , \quad (6)$$

where

- SSL_i is the computed suspended-sediment load, in tons per 15-minute interval;
- SSC_i is the computed suspended-sediment concentration for the ith value, in milligrams per liter;
- Q_i is the discharge for the ith value, in cubic feet per second; and
- C is a constant, 2.81×10^{-5}, for converting the units to tons per 15 minutes.

The computations of instantaneous (15-minute) SSL were summed to provide daily, storm, and annual loads. In cases where 15-minute unit discharge values were not available when the streamflow-gaging station was affected by ice, the mean daily discharge value was substituted. Estimated annual SSL was determined by summing the estimated 15-minute SSL transported past the streamflow-gaging station during water years 2010 and 2011 (table 7). The annual SSL was divided by the upstream drainage area to estimate the annual sediment yield. Annual sediment yields can be used to compare sediment loads among watersheds of different sizes. Daily SSLs are provided for 2010 water year in table 8 and for the 2011 water year in table 9.

For the Bakersville station, the estimated annual SSL was 17,700 tons, and the estimated yield was 464 ton/mi^2 for 11 months of the 2010 water year (table 7). Data for the 2010 water year are incomplete because data collection began on October 29, 2009. SSL was calculated for each storm with a peak discharge greater than 50 ft^3/s when unit discharge data were available. During the 2010 water year (beginning

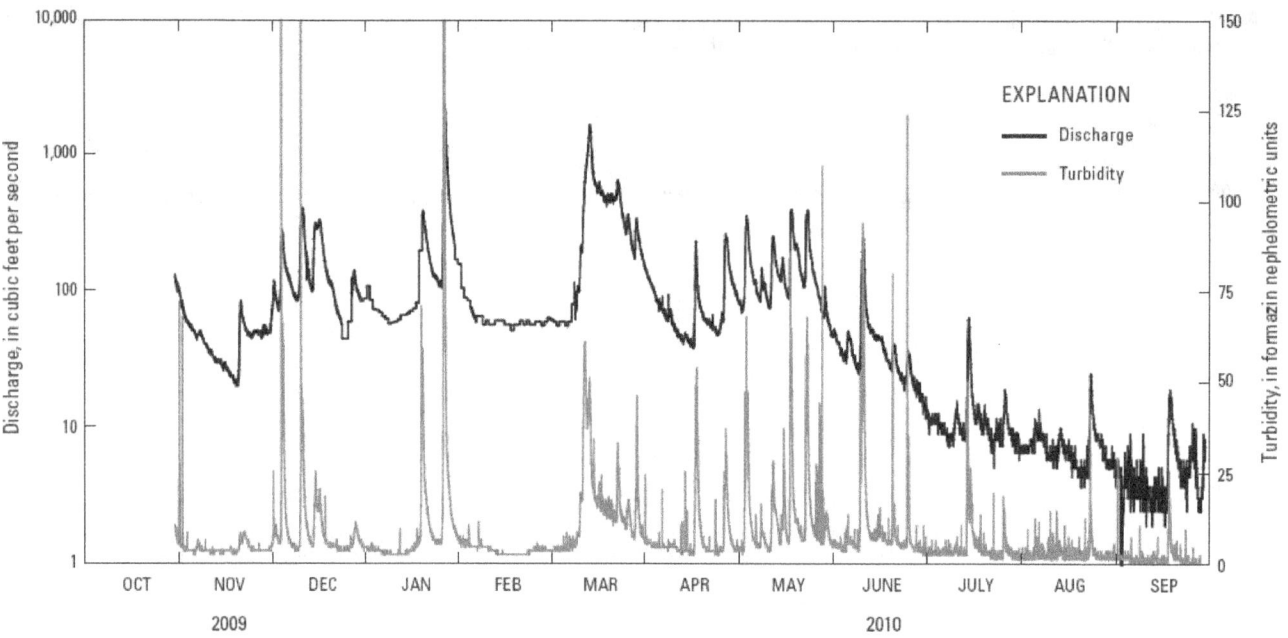

Figure 12.   Discharge and turbidity measured at station 03079600, Laurel Hill Creek near Bakersville, Pennsylvania, 2010 water year.

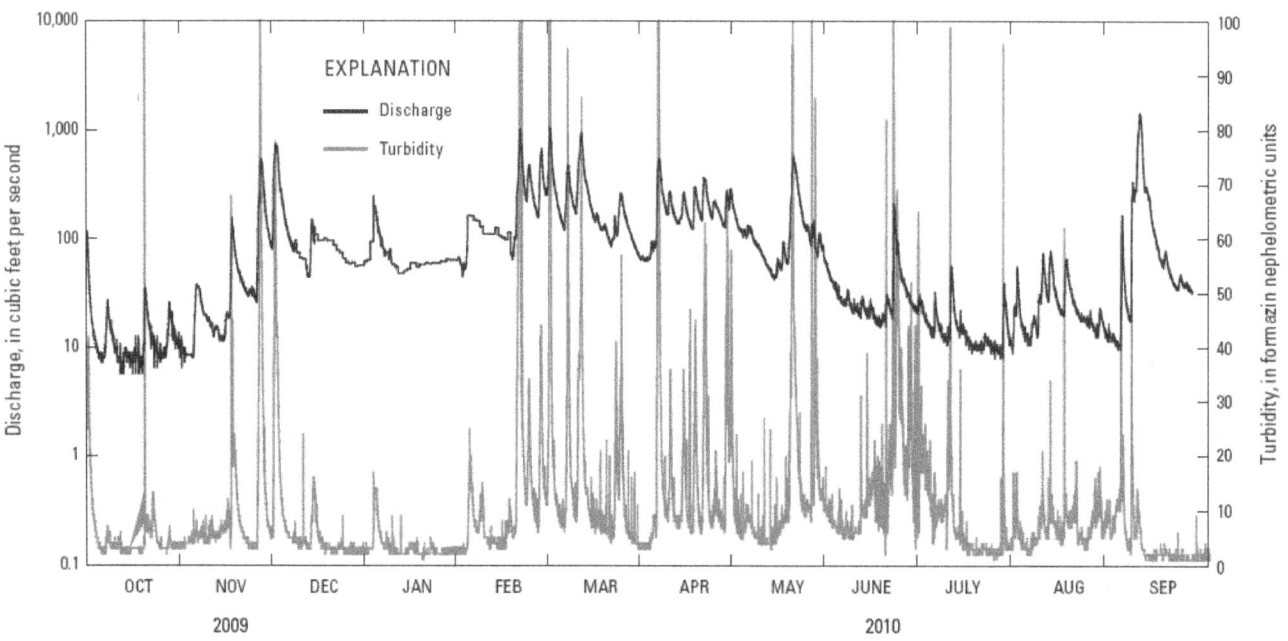

Figure 13.   Discharge and turbidity measured at station 03079600, Laurel Hill Creek near Bakersville, Pennsylvania, 2011 water year.

Table 7. Estimated annual suspended-sediment loads and yields for station 03079600, Laurel Hill Creek near Bakersville, Pennsylvania, water years 2010 and 2011.

Estimated annual load (tons per year)		Estimated annual yield (tons per square mile per year)		Percentage of sediment from stormflow	
[1]2010 water year	2011 water year	[1]2010 water year	2011 water year	[1]2010 water year	2011 water year
17,700	13,500	464	353	88.5	94.6

[1] Data collection began on October 29, 2009.

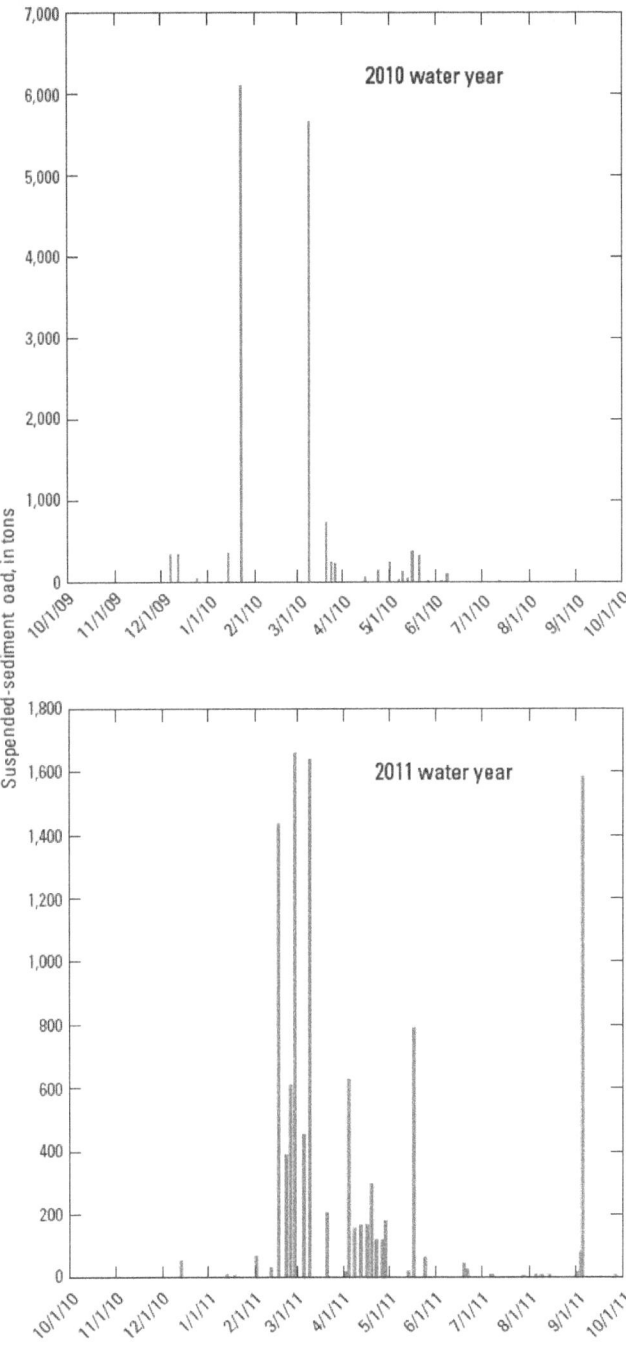

Figure 14. Estimated suspended-sediment loads from storms at station 03079600, Laurel Hill Creek near Bakersville, Pennsylvania, 2010 and 2011 water years.

October 29, 2009), there were 25 storms with peak discharges greater than 50 ft^3/s. Storms producing a discharge less than 50 ft^3/s occurred between June 1 and December 1, 2010, during a drought (Pennsylvania Department of Environmental Protection, 2012b) and during the summer of 2011. The storms were low-intensity storms that generally produced little sediment.

The storm beginning January 24, 2010, provided 34.4 percent of the annual SSL, and the storm beginning March 10, 2010, provided 31.9 percent of the annual SSL (fig. 14). Together, these two winter storms provided 66 percent of the annual SSL for the 2010 water year. The other 23 storms during the 2010 water year each provided from less than 0.01 to 4.2 percent of the annual SSL.

During the 2011 water year, there were 37 storms with peak discharges greater than 50 ft^3/s. The estimated annual SSL was 13,500 tons, and the estimated yield was 353 ton/mi^2 (table 7). During the 2011 water year, the SSLs were more evenly divided among storms than during the 2010 water year (fig. 14). Seven of 37 storms with the highest SSLs each provided a total of 65.7 percent of the annual SSL for the 2011 water year; each storm provided from 4.6 to 12.3 percent of the annual SSL. The highest cumulative SSL for the 2010 and 2011 water years generally occurred during the late winter (fig. 14). For the 2010 and 2011 water years, stormflows with the highest peak discharges generally carried the highest SSLs ($r^2 = 0.87$) (fig. 15).

Sloto and Olsen (2011) estimated sediment yields from the West Branch Brandywine and French Creek watersheds in Chester County, Pa. In comparison to the Laurel Hill Creek watershed, the French Creek watershed has about the same amount of agricultural land (33 percent) but has less forest land (44.6 percent) and more urban land (16.1 percent); the French Creek watershed had a mean sediment yield of 66.7 ton/mi^2 for the 2008 and 2009 water years. The West Branch Brandywine near Honey Brook watershed, which has a larger percentage of agricultural land (59.7 percent) than Laurel Hill Creek watershed but has much less forest land

(17.6 percent) and more urban land (13.8 percent) had a mean sediment yield of 184 (ton/mi^2)/yr for the 2008 and 2009 water years. The sediment yield in both of these watersheds is much lower than the yield for the Laurel Hill Creek watershed, which had a mean sediment yield of 409 ton/mi^2 for the 2011 water year and the last 11 months of the 2011 water year.

Table 8.   Estimated daily sediment loads at station 03079600, Laurel Hill Creek near Bakersville, Pennsylvania, 2010 water year.

[Loads are in tons; --, no data]

Day	October	November	December	January	February	March	April	May	June	July	August	September
1	--	4.7	9.4	6.4	7.5	3.1	17	5.1	2.4	0.14	0.04	0.01
2	--	3.1	6.2	4.4	7.0	3.0	13	6.1	1.6	0.12	0.04	0.01
3	--	2.7	92	4.2	4.7	2.7	10	134	1.2	0.12	0.05	0.02
4	--	2.3	40	3.9	3.6	3.0	7.5	68	1.0	0.12	0.07	0.02
5	--	2.0	17	3.2	4.1	3.2	5.7	20	2.4	0.11	0.09	0.01
6	--	2.2	11	2.8	4.1	3.0	5.2	12	1.8	0.09	0.10	0.01
7	--	1.9	7.5	2.4	3.0	3.4	4.2	7.6	1.1	0.08	0.07	0.01
8	--	1.5	6.5	2.5	3.3	6.0	3.8	14	0.8	0.07	0.04	0.01
9	--	1.2	116	2.6	2.7	6.9	4.0	11	13.2	0.09	0.04	0.01
10	--	1.1	171	2.7	2.7	17	2.8	6.22	71.2	0.18	0.04	0.01
11	--	0.93	33	3.1	2.9	90	2.2	18	7.7	0.11	0.03	0.01
12	--	0.84	14	3.2	2.6	662	1.8	64	3.4	0.08	0.07	0.01
13	--	0.81	20	3.7	2.6	1,960	1.8	25	2.8	0.17	0.08	0.01
14	--	0.73	109	4.1	2.3	2,110	1.8	16	2.5	3.9	0.05	0.01
15	--	0.72	112	4.8	2.3	487	1.5	29	2.6	0.68	0.04	0.01
16	--	0.59	65	6.5	2.0	353	2	12.4	2.1	0.19	0.03	0.13
17	--	0.51	24	49	2.3	351	48	84	1.5	0.21	0.03	0.33
18	--	0.42	13	190	2.3	264	9.29	176	1.1	0.13	0.02	0.07
19	--	1.5	11	71	2.6	231	5	52	1.2	0.15	0.02	0.03
20	--	5.4	7.3	33	2.3	249	3.3	34	1.9	0.10	0.02	0.02
21	--	3.6	4.5	19	2.7	235	2.8	17	0.88	0.07	0.02	0.01
22	--	2.5	3.2	14	2.7	262	2.6	55	0.63	0.06	0.45	0.01
23	--	2.0	1.8	12	2.8	401	2.4	198	0.53	0.08	0.19	0.00
24	--	2.1	1.8	22	3.1	152	2.0	41	1.3	0.07	0.07	0.00
25	--	2.1	3.2	4,590	3.2	86	3.1	19	1.3	0.17	0.05	0.00
26	--	2.0	13	1,270	2.7	123	51	12	0.51	0.32	0.04	0.00
27	--	2.2	15	167	3.0	56	55	9.0	0.42	0.10	0.03	0.00
28	--	2.2	8.8	63	3.3	50	17	5.7	0.34	0.07	0.03	0.04
29	--	2.3	6.4	28	--	121	10	8.3	0.25	0.06	0.03	0.03
30	11	8.5	6.7	22	--	44	6.7	3.7	0.19	0.05	0.02	13.28
31	8.0	--	9.4	10	--	25	--	2.5	--	0.04	0.02	--
Monthly total	--	65	961	6,630	91	8,350	304	1,170	130	7.9	1.9	14
Monthly yield	--	1.7	25	174	2.4	219	8.0	30.6	3.4	0.21	0.05	0.37
[1]Annual total	17,700											

[1]Data for most of October are not available.

Table 9. Estimated daily sediment loads at station 03079600, Laurel Hill Creek near Bakersville, Pennsylvania, 2011 water year.

[Loads are in tons; --, no data]

Day	October	November	December	January	February	March	April	May	June	July	August	September
1	6.9	0.10	697	7.2	3.7	620	3.5	14	1.6	0.51	0.22	10
2	0.56	0.09	194	41	32	98	4.1	13	1.2	0.33	0.16	4.4
3	0.19	0.10	48	19	31	44	7.7	13	0.91	0.74	0.18	0.63
4	0.10	0.19	20	8	21	24	15	17	0.83	0.50	0.37	13
5	0.07	1.3	11	4.9	20	18	414	13	1.1	0.34	0.28	77
6	0.10	1.2	7.0	4.5	19	207	136	8.4	0.72	0.26	1.0	245
7	0.54	0.65	6.9	4.1	12	166	47	6.8	0.64	0.29	4.5	1,038
8	0.27	0.45	5.0	2.7	10	52	39	5.3	0.67	1.8	1.5	217
9	0.14	0.38	4.0	2.2	10	41	77	4.1	0.73	1.6	3.8	43
10	0.09	0.32	3.9	1.7	10	425	30	3.0	1.0	0.39	3.8	26
11	0.06	0.26	2.3	1.7	12	911	20	2.6	0.77	0.36	1.3	11
12	0.06	0.27	6.3	1.9	8.8	165	23	2.0	0.85	0.27	0.74	5.9
13	0.09	0.20	18	2.4	8.2	77	78	2.8	0.85	0.21	0.54	4.2
14	0.07	0.22	13	3.1	9.0	37	38	3.8	0.57	0.14	3.1	2.8
15	0.07	0.53	8.4	2.7	14	23	20	6.7	0.52	0.12	3.3	3.2
16	0.09	2.7	7.3	2.4	5.3	24	40	5.6	0.55	0.11	1.4	2.6
17	0.08	20	7.8	2.4	27	17	95	16	0.82	0.11	0.96	1.51
18	0.06	6.4	7.1	2.4	596	16	32	301	1.2	0.11	0.68	1.04
19	0.92	2.9	7.0	2.6	713	14	73	322	0.92	0.15	0.45	0.89
20	0.59	1.8	5.8	2.8	88	8.7	165	107	39	0.13	0.38	1.2
21	0.26	1.2	4.9	2.9	170	9.0	47	41	16	0.12	0.35	0.98
22	0.19	0.98	4.6	2.6	144	24	28	20	5.6	0.10	0.42	0.87
23	0.12	1.1	3.4	2.6	51	44	47	18	3.0	0.10	0.38	0.74
24	0.09	0.83	2.8	2.7	28	82	36	15	1.9	0.10	0.33	0.68
25	0.08	31	2.9	2.8	313	30	24	20	1.8	0.94	0.58	0.60
26	0.11	370	2.9	3.2	234	16	28	11	1.1	0.68	0.47	0.52
27	0.44	111	2.4	3.1	77	11	98	14	1.1	0.20	0.29	1.7
28	0.31	20	2.5	3.2	854	7.1	96	8.2	1.5	0.16	0.26	1.7
29	0.18	8.8	2.5	3.2	--	5.0	46	4.6	1.2	0.66	0.22	1.1
30	0.14	182	2.9	3.2	--	3.9	22	3.0	0.70	1.9	0.22	0.91
31	0.11	--	3.3	2.1	--	3.6	--	2.1	--	0.38	0.21	--
Monthly total	13	768	1,110	151	3,520	3,221	1,830	1,030	89	14	32	1,720
Monthly yield	0.3	20	29	4.0	92	84	48	27	2.3	0.36	0.85	45
Annual total	13,500											

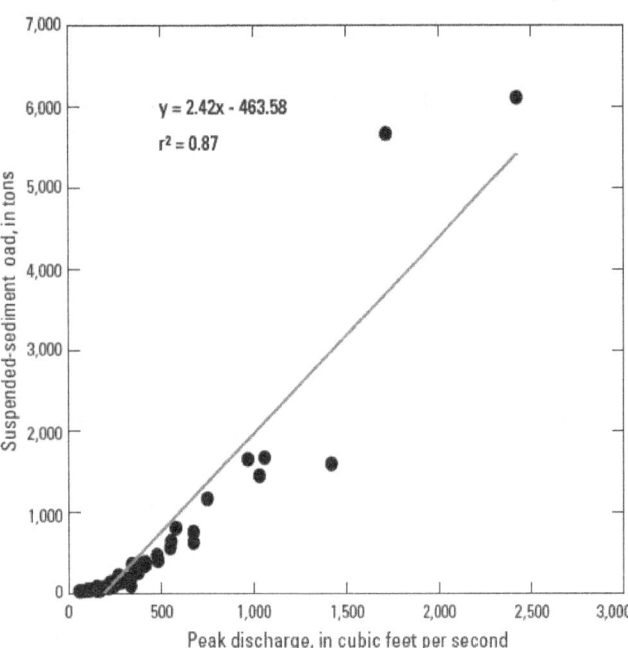

Figure 15.   Relation of estimated storm suspended-sediment load to peak discharge at station 03079600, Laurel Hill Creek near Bakersville, Pennsylvania, 2010 and 2011 water years.

Sediment-Source Assessment Using Sediment Fingerprints

At the Bakersville station, 10 suspended-sediment samples were collected during 6 storms for sediment-source analysis (fig. 16; table 10). Except for the March 12, 2010, sample, sediment collection occurred over several hours during the time periods shown on the flood hydrographs in figure 16. Ninety samples were collected from five source areas: 20 from cropland, 9 from pasture, 18 from forest land, 20 from unpaved roads, and 23 from streambanks (fig. 7; appendix 5). Nine streambank samples were collected in first-order channels, 9 in second-order channels, 2 in third-order channels, and 3 in fourth-order channels (fig. 7; appendix 5). The percentage of fines (silt and clay) in the source samples for the Laurel Hill Creek watershed ranged from 2 to 74 percent; streambank samples had the highest percentage of fines, and unpaved road samples had the lowest percentage (appendix 5).

Statistical Methods

Several analytical and statistical steps were used to determine which tracers were most appropriate in defining sediment sources as follows: (1) removing outliers in each source type, (2) bracketing the fluvial samples by the source type, (3)

performing stepwise discriminant function analysis (DFA), (4) evaluating the capacity of the set of characteristic variables to discriminate between source types, and (5) identifying source percentages using an "unmixing" model (fig. 17).

The first step in the statistical procedure was to remove outliers. The presence of outliers can lead to errors in data analysis and statistical conclusions (Helsel and Hirsch, 1992). Before the outlier test was performed, tracers for each source type were tested for a normal distribution using the Shapiro-Wilk test (null hypothesis is that samples are random and come from a normal distribution). All variables that were not normally distributed at a 95-percent confidence interval were tested again for normality after transformation using one of the following: log, power, square root, cube root, or inverse function (Helsel and Hirsch, 1992). The best transformation for normality was selected, and the three standard deviation rule (Wainer, 1976) was applied to that data set. Tracers in each source type that were greater or less than three times the standard deviation of the average value of a tracer in that source type were considered outliers, and the entire sample of that tracer was removed.

A requirement of sediment fingerprinting is that the fluvial tracers must be conservative and not change during transport from the source to the sampling point. Consequently, the next step in the statistical analysis was to determine that for a given tracer, the fluvial samples were bracketed by the sources within the range of error for each tracer. Any tracers that did not satisfy this constraint within measurement error were considered to be non-conservative and were removed from further consideration.

Collins and Walling (2002) and Collins and others (1997) have suggested that a composite of several tracers provides a greater ability to discriminate between sources than a single tracer. To create the optimal group of tracers, a stepwise DFA was used to select tracers. For this procedure, normality among the variables being analyzed was assumed; therefore, all variables used in the DFA were tested for normality using the Shapiro-Wilk test. All variables that were not normally distributed at a 95-percent confidence interval were tested again for normality after transformation using either a log, power, square root, cube root, or inverse function.

The best transformation for normality was selected, and stepwise DFA was performed on the data. Stepwise DFA incrementally identifies the tracers that significantly contribute to correctly identifying the sediment sources and rejects variables that do not contribute based on the minimization of the computed value of the variable Wilks' lambda (Collins and others, 1997). A lambda close to 1.0 indicates that the means of all tracers chosen are equal, and differences among groups cannot be distinguished. A lambda close to zero occurs when any two groups are well separated (within group variability is small compared to overall variability). Thus, the model selects a combination of tracers that provide optimal separation; no better separation can be achieved using fewer or more tracers. A significance value of 0.05 was used in the procedure. The statistical program SAS was used in the stepwise DFA.

Table 10. Description of streamflows sampled during storms for sediment fingerprinting at station 03079600, Laurel Hill Creek near Bakersville, Pennsylvania, March 2010 to April 2011.

[ft³/s, cubic feet per second]

Date	Sample start (date: hours and minutes)	Sample end (date: hours and minutes)	Range of sampled flows (ft³/s)	Peak discharge of sampled event (ft³/s)	Peak flow (date: hours and minutes)	Part of storm hydrograph during which sample was collected	Suspended-sediment load for days on which the sediment sample was collected (tons/day)
3/12/2010	3/12/2010:0800	3/12/2010:0941	600–625	1,720	3/14/2010:0045	Rising	662
3/12/2010	3/12/2010:0942	3/12/2010:1105	626–652	1,720	3/14/2010:0045	Rising	662
5/17–18/2010	5/17/2010:1700	5/18/2010:0400	250–408	408	5/18/2010:0430	Rising	261
5/17–18/2010	5/18/2010:0500	5/18/2010:1600	293–395	408	5/18/2010:0430	Rising	28.8
9/30–10/1/2010	9/30/2010:1230	9/30/2010:1830	19–142	148	9/30/2010:1930	Rising	13.3
9/30–10/1/2010	9/30/2010:1930	10/01/2010:1230	45–148	148	9/30/2010:1930	Rising	20.1
11/16–17/2010	11/16/2010:2030	11/17/2010:1230	58–154	154	11/17/2010: 0445	Rising and falling	20.2
4/5/2011	4/5/2011:0500	4/5/2011:1200	413–526	552	4/5/2011:1445	Rising	414
4/5/2011	4/5/2011:1215	4/5/2011:2000	515–552	552	4/5/2011:1445	Rising and falling	414
4/19–20/2011	4/19/2011:2000	4/20/2011:0900	338–368	368	4/19/2011:2200	Rising and falling	238

Another requirement of sediment fingerprinting is that tracers have a unique value for each source. The Mahalanobis distance statistic was used to verify that the set of normalized tracers determined from the stepwise DFA can correctly identify each source type. The Mahalanobis distance statistic tests the distance between source types for a set of tracers (Rao, 1965). A probability value of 0.05 was used in the Mahalanobis distance statistic test. The Mahalanobis distance statistic was run on the set of normalized tracers determined from the stepwise DFA. If no tracer is found to differentiate between two sources, a decision is made to combine the sources into one source type, for example, cropland and pasture into agriculture. Stepwise DFA and the Mahalanobis distance statistic tests are then repeated on the new data set (fig. 17). The Mahalanobis distance statistic test has been used in other sediment studies to assist in discriminating sources (Karlin, 1980; Minella and others, 2008).

The final step in the statistical analysis was to determine the significant sources of sediment using an "unmixing model." The tracer values that were determined from the stepwise DFA are used in the "unmixing" model but in their original (non-transformed) form. A modified version of the "unmixing" equation of Collins and others (2010) was used to determine source percentages:

$$\left(\sum_{i=1}^{n} \{ (C_i - (\sum_{s=1}^{m} P_s S_{si})) / C_i \}^2 W_i \right) , \tag{7}$$

and

$$\sum_{i=1}^{n} = P_S = 1 , \tag{8}$$

where

C_i	is the concentration of tracer property (i) in the suspended sediment collected during storms,
P_s	is the optimized percentage contribution from the source type(s),
S_{si}	is the mean concentration of tracer property (i) in source type(s),
W_i	is the tracer discriminatory weighting,
n	is the number of fingerprint properties comprising the optimum composite fingerprint, and
m	is the number of sediment-source types.

The "unmixing" model iteratively tests for the lowest error value using all possible source percentage combinations. A step of 0.01 is used in the source computations. In a five-source model, there are 4,598,126 iterations to find the lowest

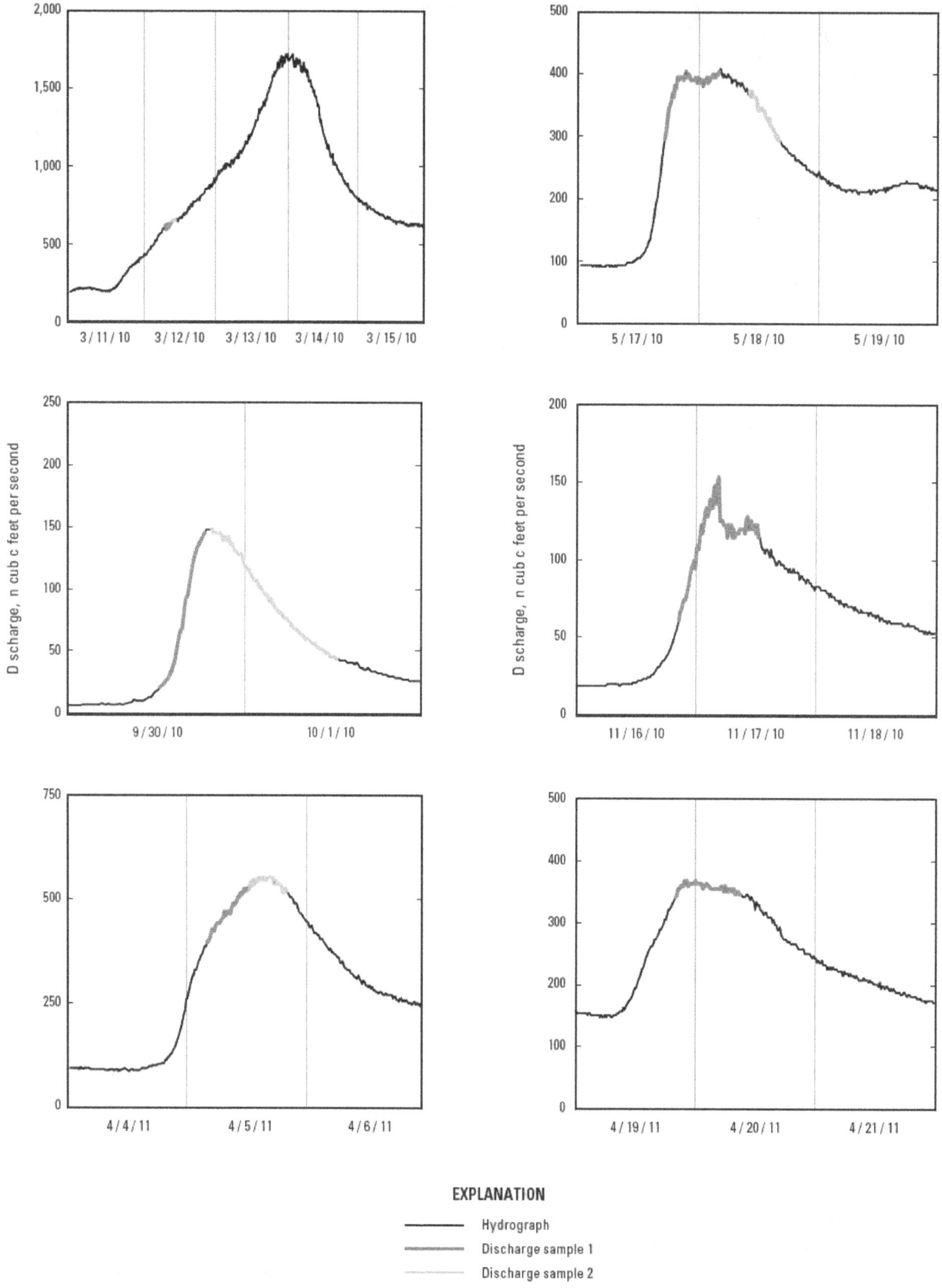

Figure 16.   Discharge and suspended-sediment sample collection during storms for source analysis at station 03079600, Laurel Hill Creek near Bakersville, Somerset County, Pennsylvania.

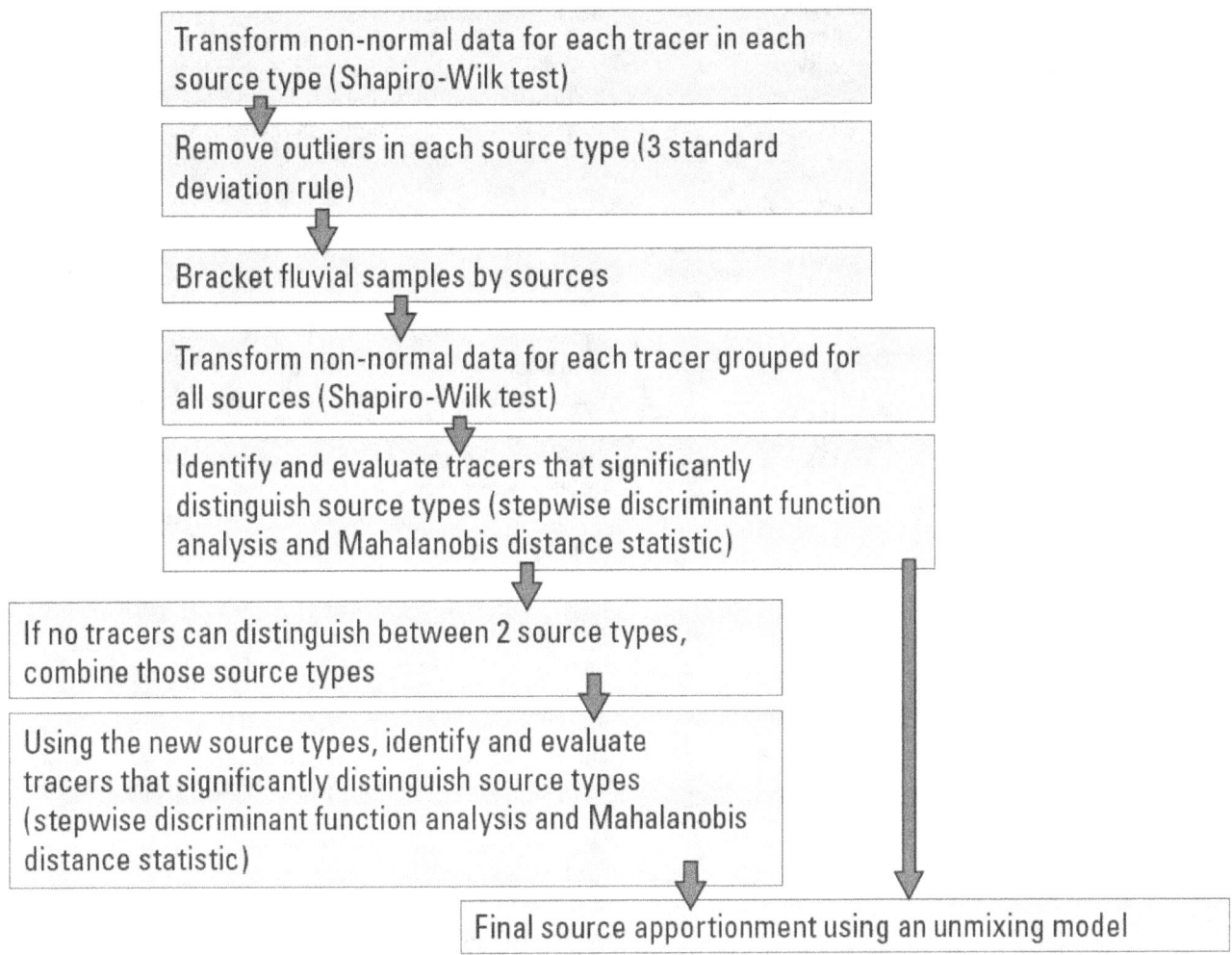

Figure 17. Flow chart of steps used to determine the significant sources of sediment in the Laurel Hill Creek watershed, Somerset County, Pennsylvania.

error value. Wi, the tracer discriminatory weighting value, is a weighting used to reflect tracer discriminatory power (eq. 7). This weighting is based on the relative discriminatory power of each individual tracer provided by the results of the stepwise DFA.

Examination for outliers in the source types showed that no tracers were outside the specified range (greater or less than three times the standard deviation from the average) for that source type. Examination of the fluvial tracers in relation to the source samples showed that the measured values for eight tracers (Ba, Cd, Co, Mg, Mn, Na, Ni, Zn) were outside the range of the measured source values and were removed (appendix 5). The results of the stepwise DFA on the final 31 tracers indicated that 11 tracers were significant (Bi, Nb, Y, P, Fe, Ti, Pb, Cu, Th, Cs, K). Results from the Mahalanobis

distance test on the 11 tracers determined from the stepwise DFA indicated that the composite set of all 11 tracers was not able to distinguish between pasture and cropland (p=0.193), and these source types were combined into one source group called agricultural land. On the basis of conversations with landowners in the Laurel Hill Creek watershed, many of the pasture fields were cropland in the recent past or are rotated into cropland every few years, and thus, combining cropland and pasture into one source group was representative of agriculture in the watershed.

Stepwise DFA was run again on the final 31 tracers for the four sediment source groups (agricultural lands, streambanks, unpaved roads, and forests) and showed that 11 tracers were significant (Bi, Nb, Y, P, Fe, Ti, Pb, Cu, Th, Cs, K) (table 11). The cumulative percentage of source-type samples

Table 11. The optimum composite set of tracers for discriminating individual sediment-source types in the Laurel Hill Creek watershed, Somerset County, Pennsylvania.

[Bi, Bismuth; Nb, Niobium; Y, Yttrium; P, Phosphorus; Fe, Iron; Ti, Titanium; Pb, Lead; Cu, Copper; Th, Thorium; Cs, Cesium; K, Potassium]

Step	Tracer selected	Cumulative percentage of source-type samples classified correctly	Percentage of source-type samples classified correctly	Tracer discriminatory weighting value
1	Bi	66.4	33.6	2.2
2	Nb	73.3	41.9	1.9
3	Y	89.7	42.8	1.9
4	P	94.2	47.7	1.7
5	Fe	94.2	47.4	1.7
6	Ti	98.9	42.9	1.9
7	Pb	98.9	38.2	2.1
8	Cu	98.9	45.8	1.8
9	Th	98.9	69.9	1.0
10	Cs	97.8	63.6	1.2
11	K	97.8	63.6	1.2

correctly classified was 97.8 percent (table 11). Results from the Mahalanobis distance test on the 11 tracers determined from the stepwise DFA indicate that the composite set of all 11 tracers was able to distinguish between individual source types (table 11); 4 source types were used in the statistical analysis of sediment-source types (table 12).

Table 12. Probability that the 11 tracers from the stepwise discriminant function analysis can distinguish between individual source types using the Mahalanobis distance statistic, Laurel Hill Creek watershed, Somerset County, Pennsylvania.

[Values less than 0.05 indicate that the composite set of tracers can be used to distinguish between the source types]

	Streambanks	Agricultural lands	Forests	Unpaved roads
Streambanks	1	<0.0001	<0.0001	<0.0001
Agricultural lands		1	<0.0001	<0.0001
Forests			1	<0.0001
Unpaved roads				1

Sediment Sources

The "unmixing" model was applied to the final 11 tracers for the 10 stormflow samples (table 13). Averaging the source percentages for the 10 storms showed that agricultural lands were the dominant source of fine-grained sediment (53 percent). Streambanks contributed 30 percent; unpaved roads, 17 percent; and forests, 0 percent.

Sediment flux is highly correlated with discharge. The discharge at the time of suspended-sediment sample collection is an indicator of the energy that is available to erode and transport sediment. When the relation between source contributions and discharge were plotted, the graph showed that agricultural lands were an important sediment source during high discharges. Samples collected during the highest discharges also had the highest suspended-sediment loads on that day. Streambanks and unpaved roads were a source of sediment over the range of discharges and sediment-transport conditions (fig. 18).

For discharges of 19 to 154 ft³/s, streambanks were the major contributor of sediment (average of 47 percent), followed by unpaved roads (average of 39 percent), and agricultural lands (average of 13 percent). For discharges of 250 to 552 ft³/s, agricultural lands were the major contributor of sediment (average of 73 percent) followed by streambanks (average of 17 percent), and unpaved roads (average of 10 percent). For discharges of 600 to 652 ft³/s, agricultural lands were the major contributor of sediment (average of 64 percent), followed by streambanks (average of 33 percent), and unpaved roads (average of 3 percent).

The highest contribution of sediment from streambanks (61 percent) was observed for the second lowest sampled discharges (45 to 148 ft³/s on September 30 to October 1, 2010), which had low daily suspended-sediment loads (fig. 18). The highest contribution from unpaved roads (58 percent) was observed for the third lowest sampled discharges (58 to 154 ft³/s on November 16–17, 2010), which had low daily suspended-sediment loads (fig. 18).

Table 13. Sediment sources determined using the unmixing model and 15 tracers for samples collected during 6 storms, Laurel Hill Creek watershed, Somerset County, Pennsylvania, 2010–11.

Date	Sample start (date: hours and minutes)	Sample end (date: hours and minutes)	Agricultural land (percent)	Forest (percent)	Unpaved road (percent)	Streambank (percent)	Error
3/12/2010	3/12/2010:0800	3/12/2010:0941	62	0	3	35	0.34
3/12/2010	3/12/2010:0942	3/12/2010:1105	65	0	3	32	0.39
5/17–18/2010	5/17/2010:1700	5/18/2010:0400	66	0	0	34	0.35
5/17–18/2010	5/18/2010:0500	5/18/2010:1600	75	0	0	25	0.32
9/30–10/1/2010	9/30/2010:1230	9/30/2010:1830	16	0	45	39	0.69
9/30–10/1/2010	9/30/2010:1930	10/01/2010:1230	24	0	15	61	0.56
11/16–17/2010	11/16/2010:2030	11/17/2010:1230	0	0	58	42	1.11
4/5/2011	4/5/2011:0500	4/5/2011:1200	76	0	17	7	0.21
4/5/2011	4/5/2011:1215	4/5/2011:2000	78	0	18	4	0.32
4/19–20/2011	4/19/2011:2000	4/20/2011:0900	71	0	14	15	0.41
		Average	53	0	17	30	

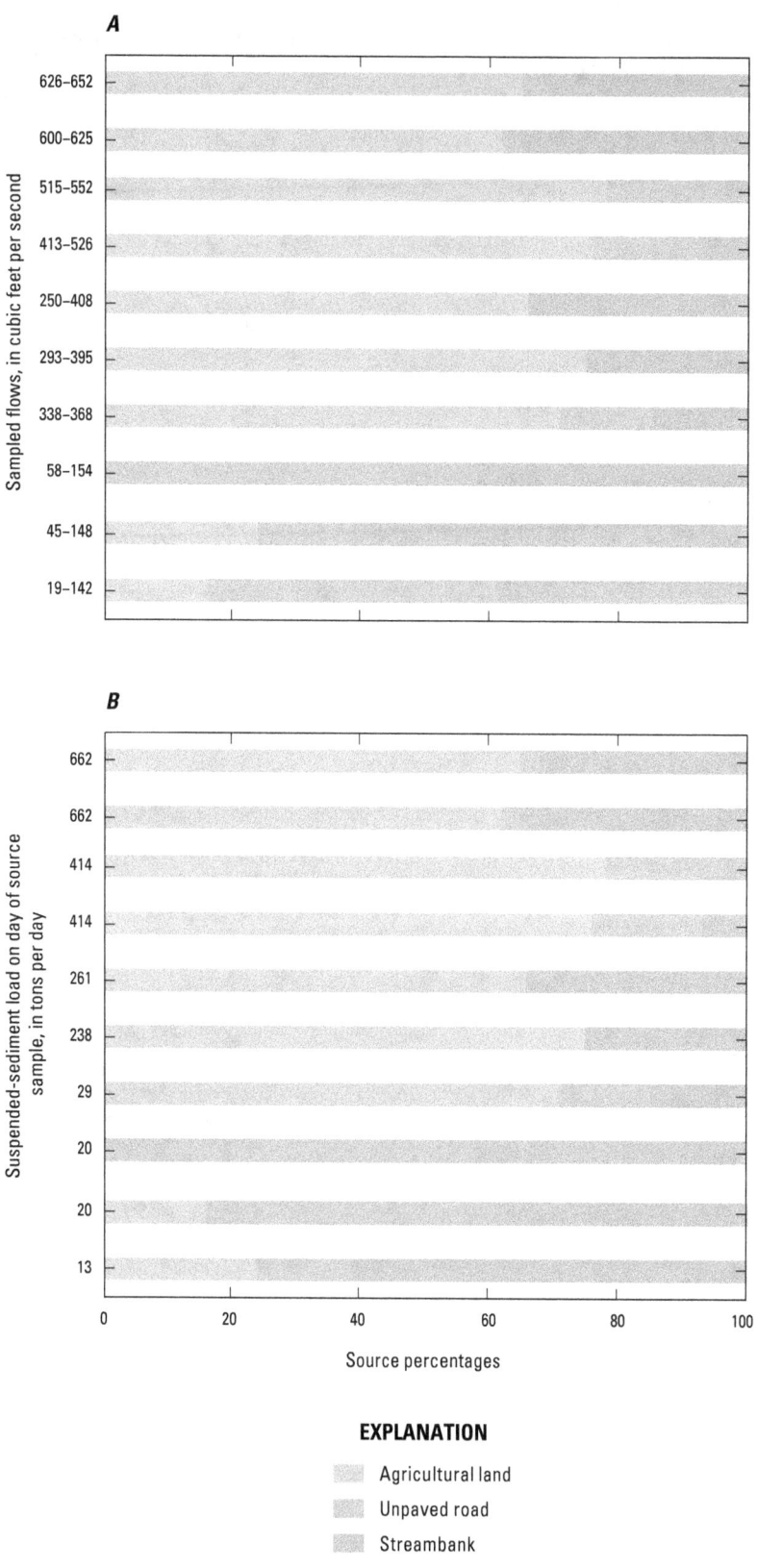

Figure 18. Sediment sources by *A*, a range of discharges during sampling and *B*, suspended-sediment load on the day of sample collection, at station 03079600, Laurel Hill Creek near Bakersville, Pennsylvania.

Summary and Conclusions

Laurel Hill Creek is a 125 mi^2 watershed located mostly in Somerset County, Pennsylvania, with small areas extending into Fayette and Westmoreland Counties. On the basis of land-use data for 2001, the Laurel Hill Creek watershed is 63.4 percent forest, 27.2 percent agricultural, 4.9 percent residential, 3.0 percent wetlands and open water, and 1.6 percent commercial/industrial and mining. Withdrawals are made from surface-water and groundwater sources to supply multiple users. The upper part of the Laurel Hill Creek watershed, upstream from the U.S. Geological Survey (USGS) stream-flow-gaging station near Bakersville, is on the PaDEP 303(d) list of impaired streams because of sedimentation (siltation), elevated nutrient concentrations, and low dissolved oxygen concentrations. The objectives of this study in the Laurel Hill Creek watershed were to (1) estimate the annual nitrogen load, (2) estimate the annual sediment load, and (3) identify the major sources of fine-grained sediment using the sediment-fingerprinting approach. This study was done by the USGS in cooperation with the Somerset County Conservation District.

Discharge data were collected at two streamflow-gaging stations on Laurel Hill Creek: Laurel Hill Creek near Bakersville, Pa., (station 03079600) and Laurel Hill Creek at Ursina, Pa., (station 03080000). Turbidity data were collected at the Bakersville station at 15-minute intervals using a turbidity sensor. Suspended-sediment samples were collected for base-flow and storm events at the Bakersville, Ursina, and Trent (Laurel Hill Creek below Laurel Hill Creek Lake at Trent, station 03079655) stations. Suspended-sediment and nutrient samples were collected at the Bakersville station during high (storm) flows by an automatic sampler. Water samples were collected manually for nutrient analysis at the Ursina and Trent sites.

Water samples for nutrient analysis were collected at the Bakersville and Ursina stations from July 2009 through September 2011. Concentrations of nutrients, in general, were low. Most concentrations of phosphorous were less than the detection limit of 0.02 milligrams per liter (mg/L). Most water samples had concentrations of nitrate plus nitrite of less than 1 mg/L; only one sample exceeded a concentration of 1 mg/L.

At the Bakersville station, concentrations of dissolved total nitrogen ranged from 0.61 to 1.3 mg/L in base-flow samples, 1.1 to 1.3 mg/L in stormflow grab samples, and 0.52 to 1.1 mg/L in stormflow composite samples. Median concentrations for base-flow and stormflow composite samples were similar at 0.88 and 0.91 mg/L, respectively. The median concentration for storm grab samples was 1.2 mg/L. Concentrations of total nitrogen ranged from 0.63 to 1.3 mg/L for base-flow samples, 1.1 to 1.4 mg/L for stormflow grab samples, and 0.57 to 1.5 mg/L for stormflow composite samples. Median concentrations were 0.88 mg/L for base-flow samples, 1.2 mg/L for stormflow grab samples, and 1.2 mg/L for stormflow composite samples.

At the Ursina station, concentrations of dissolved total nitrogen ranged from 0.12 to 1.1 mg/L for base-flow samples and 0.52 to 0.97 mg/L for stormflow samples. Median concentrations for base-flow and stormflow samples were similar at 0.58 and 0.70 mg/L, respectively. Concentrations of total nitrogen ranged from 0.25 to 0.92 mg/L for base-flow samples, and the median concentration was 0.57 mg/L.

Monthly total nitrogen loads were estimated for the Bakersville station. The estimated total nitrogen load was 262 pounds (lb) for 11 months of the 2010 water year (November 2009 to September 2010) and 266 lb for the 2011 water year. Estimated mean monthly total nitrogen loads were similar for both water years. The estimated mean monthly total nitrogen load was 23.9 pounds per month (lb/mo) for the 2010 water year and 22.2 lb/mo for the 2011 water year. Most of the total nitrogen load was from stormflows. The estimated mean monthly load in stormflow was 18.3 lb/mo for the 2010 water year and 17.9 lb/mo for the 2011 water year. The load in stormflow made up 76.6 percent of the total load for the 2010 water year and 80.6 percent of the total load for the 2011 water year. The estimated monthly total nitrogen loads were higher during the winter and spring (December through May) than during the summer (June through August).

For the Bakersville station, the estimated annual suspended-sediment load (SSL) was 17,700 tons, and the estimated yield was 464 tons per square mile (ton/mi^2) for 11 months of the 2010 water year (November 2009 to September 2010). During the 2010 water year (beginning October 29, 2009), there were 25 storms with peak discharges greater than 50 ft^3/s. The storm beginning January 24, 2010, provided 34.4 percent of the annual SSL, and the storm beginning March 10, 2010, provided 31.9 percent of the annual SSL. Together, these two winter storms provided 66 percent of the annual SSL for the 2010 water year. The other 23 storms during the 2010 water year each provided from less than 0.01 to 4.2 percent of the annual SSL.

During the 2011 water year, there were 37 storms with peak discharges greater than 50 ft^3/s. The estimated annual SSL was 13,500 tons, and the estimated yield was 353 ton/mi^2. For the 2011 water year, the SSLs were more evenly divided among storms than for the 2010 water year. Seven of 37 storms with the highest SSLs provided a total of 65.7 percent of the annual SSL for the 2011 water year; each storm provided from 4.6 to 12.3 percent of the annual SSL. The highest cumulative SSL for the 2010 and 2011 water years generally occurred during the late winter. For the 2010 and 2011 water years, storms with the highest peak discharges generally carried the highest SSL.

The sediment-fingerprinting approach was used to identify sources of fine-grained suspended sediment. The approach entailed the identification of specific sediment sources through the establishment of a set of tracers that uniquely defined each source in the watershed. Source-sediment samples were collected upstream from the Bakersville station at upland source areas and streambanks. Sediment sources were identified as agricultural lands (cropland and pasture), forests, unpaved roads, and streambanks.

At the Bakersville station, 10 suspended-sediment samples were collected during 6 storms for sediment-source analysis. Statistical analysis determined that pasture and cropland could not be discriminated by the set of tracers and were combined into one source group—agricultural lands. The four sediment sources in the Laurel Hill Creek watershed (agricultural lands, unpaved roads, streambanks, and forests) were differentiated by 11 tracers. An "unmixing" model applied to the 11 tracers showed that agricultural lands were the major source of sediment, contributing an average of 53 percent of the sediment in the 10 storm samples. Contributions from streambanks, unpaved roads, and forests for the 10 storm samples averaged 30, 17, and 0 percent, respectively. Agricultural lands were the major contributor of sediment during the highest sampled discharges. The highest discharges also produced the highest total nitrogen and suspended sediment loads.

For the lowest sampled discharges of 19 to 154 ft³/s, streambanks were the major contributor of sediment (average of 47 percent), followed by unpaved roads (average of 39 percent), and agricultural lands (average of 13 percent). For discharges of 250 to 552 ft³/s, agricultural lands were the major contributor of sediment (average of 73 percent), followed by streambanks (average of 17 percent), and unpaved roads (average of 10 percent). For the highest sampled discharges of 600 and 652 ft³/s, agricultural lands were the major contributor of sediment (average of 64 percent), followed by streambanks (average of 33 percent), and unpaved roads (average of 3 percent).

References Cited

American Society for Testing and Materials, 2002, Standard test methods for determining sediment concentration in water samples: ASTM Book of Standards: West Conshohocken, Pa., American Society for Testing and Materials.

Bragg, J.M., Sobieszczyk, Steven, Uhrich, M.A., and Piatt, D.R., 2007, Suspended-sediment loads and yields in the North Santiam River Basin, Oregon, water years 1999–2004: U.S. Geological Survey Scientific Investigations Report 2007–5187, 26 p. (Also available at *http://pubs.usgs.gov/sir/2007/5187/*.)

Brenna, J.T., Corso, T.N., Tobias, H.J., and Caimi, R.J., 1997, High-precision continuous-flow isotope-ratio mass spectrometry: Mass Spectrometry Reviews, v. 16, p. 227–258.

Collins, A.L., and Walling, D.E., 2002, Selecting fingerprint properties for discriminating potential suspended sediment sources in river basins: Journal of Hydrology, v. 261, p. 218–244.

Collins, A.L., Walling, D.E., and Leeks, G.J.L., 1997, Sediment sources in the Upper Severn catchment—a fingerprinting approach: Hydrology and Earth System Sciences, v. 1, p. 509–521.

Collins, A.L., Walling, D.E., Webb, L., and King, P., 2010, Apportioning catchment scale sediment sources using a modified composite fingerprinting technique incorporating property weightings and prior information: Geoderma, v. 155, p. 249–261.

Devereux, O.H., Prestegaard, K.L., Needelman, B.A., and Gellis, A.C., 2010, Suspended-sediment sources in an urban watershed, Northeast Branch Anacostia River, Maryland: Hydrological Processes, v. 24, no. 11, p. 1391–1403.

Duan, Naihua, 1983, Smearing estimate—a nonparametric retransformation method: Journal of the American Statistical Association, v. 78, no. 383, p. 605–610.

FTS, Inc., 2011, Frequently-asked questions DTS 12, accessed November 29, 2011, at *http://www.ftshydrology.com/Products/Sensors/DTS-12/faqs.html.*

Gellis, A.C., Hupp, C.R., Pavich, M.J., Landwehr, J.M., Banks, W.S.L., Hubbard, B.E., Langland, M.J., Ritchie, J.C., and Reuter, J.M., 2009, Sources, transport, and storage of sediment at selected sites in the Chesapeake Bay Watershed: U.S. Geological Survey Scientific Investigations Report 2008–5186, 95 p. (Also available at *http://pubs.usgs.gov/sir/2008/5186/pdf/sir2008-5186rev1142011.pdf.*)

Gellis, A.C., and Walling, D.E., 2011, Sediment-source fingerprinting (tracing) and sediment budgets as tools in targeting river and watershed restoration programs, *in* Simon, A., Bennett, S., Castro, J.M., eds., Stream restoration in dynamic fluvial systems: Scientific Approaches, Analyses, and Tools, American Geophysical Union Monograph Series 194, p. 263–291.

Harris, D., Horwath, W.R., and van Kessel, C., 2001, Acid fumigation of soils to remove carbonates prior to total organic carbon or carbon-13 isotopic analysis: Journal of the Soil Science Society of America, v. 65, p. 1853–1856.

Helsel, D.R., and Hirsch, R.M., 1992, Statistical methods in water resources: Amsterdam, Elsevier Science B.V., 529 p.

Helsel, D.R., and Hirsch, R.M., 2002, Statistical methods in water resources: U.S. Geological Survey Techniques of Water-Resources Investigations, chap. A3, book 4, 522 p. (Also available at *http://pubs.usgs.gov/twri/twri4a3/.*)

Kammerer, P.A., Garn, H.S., Rasmussen, P.W., and Ball, J.R., 1998, A comparison of water-quality sample collection methods used by the U.S. Geological Survey and the Wisconsin Department of Natural Resources: Proceedings of the National Water-Quality Monitoring Council, National Conference on Monitoring, Reno, Nevada, July 7–9, 1998: Washington, D.C., U.S. Environmental Protection Agency, p. III–259–269.

Karlin, R., 1980, Sediment sources and clay mineral distributions off the Oregon coast: Journal of Sedimentary Petrology, v. 50, no. 2, p. 543–559.

Larsen, M.C., Gellis, A.C., Glysson, G.D., Gray, J.R., and Horowitz, A.J., 2010, Fluvial sediment in the environment: a national problem: Proceedings, 2nd Joint Federal Interagency Conference, Las Vegas, Nev., June 27–July 1, 2010, 15 p., accessed March 10, 2012, at *http://acwi.gov/ sos/pubs/2ndJFIC/Contents/OS_Larsen_9fisc_sediment_ vision_3_4_2010.pdf.*

Lee, C.J., Rasmussen, P.P., and Ziegler, A.C., 2008, Characterization of suspended-sediment loading to and from John Redmond Reservoir, east-central Kansas, 2007–08: U.S. Geological Survey Scientific Investigations Report 2008–5123, 25 p. (Also available at *http://pubs.usgs.gov/ sir/2008/5123/.*)

Miles, C.E., and Whitfield, T.G., compilers, 2001, Bedrock geology of Pennsylvania: Pennsylvania Geological Survey, 4th ser., digital dataset, scale 1:250,000.

Minella, J.P.G.,Walling, D.E., and Merten, G.H., 2008, Combining traditional monitoring and sediment source tracing techniques to assess the impact of improved land management on catchment sediment yields: Journal of Hydrology, v. 348, p. 546–563.

Motha, J.A., Wallbrink, P.J., Hairsine, P.B., and Grayson, R.B., 2003, Determining the sources of suspended sediment in a forested catchment in southwestern Australia: Water Resources Research, v. 39, no. 3, p. 1056–1069.

Nagle, G.N., Fahey, T.J., Ritchie, J.C., and Woodbury, P.B., 2007, Variations in sediment sources and yields in the Finger Lakes and Catskills regions of New York: Hydrological Processes, v. 21, no. 6, p. 828–838.

National Oceanic and Atmospheric Administration, 2001, Monthly station normals of temperature, precipitation, and heating and cooling degree days 1971–2000: Climatography of the United States, no. 81, 30 p.

Papanicolaou, A.N., Fox, J.F., and Marshall, J., 2003, Soil fingerprinting in the Palouse Basin, USA using stable carbon and nitrogen isotopes: International Journal of Sediment Research, v. 18, no. 2, p. 278–284.

Pennsylvania Department of Environmental Protection, 2006a, Guidelines for identification of Critical Water Planning Areas (PADEP Guidance Document); Harrisburg, Pa., Pennsylvania Department of Environmental Protection, report number 392-2130-014, 17 p.

Pennsylvania Department of Environmental Protection, 2006b, Pennsylvania's Surface Water Quality Monitoring Network: Harrisburg, Pa., Pennsylvania Department of Environmental Protection, Bureau of Water Standards and Facility Regulation Report 3800-BK-DEP0636, 103 p.

Pennsylvania Department of Environmental Protection, 2009, Supporting documentation Laurel Hill Creek, Somerset and Fayette Counties nomination for critical water planning area under Pennsylvania state water plan, accessed February 24, 2012, at *http://www.pawaterplan.dep.state.pa.us/ docs/TechnicalDocuments/SupportingDocumentation/Laurel_Hill_Report.pdf.*

Pennsylvania Department of Environmental Protection, 2012a, 2010 Pennsylvania integrated water quality monitoring and assessment report—streams, category 5 waterbodies, pollutants requiring a TMDL: accessed February 23, 2012, at *http://files.dep.state.pa.us/Water/Drinking%20Water%20 and%20Facility%.*

Pennsylvania Department of Environmental Protection, 2012b, Drought status map history, accessed June 7, 2012, at *http:// www.portal.state.pa.us/portal/server.pt?open=514&objID= 554262&mode=2.*

Pennsylvania Department of Transportation, 2009a, PennDOT —Pennsylvania state roads 2009: Pennsylvania Department of Transportation, Bureau of Planning and Research, Cartographic Information Division, digital dataset, accessed October 9, 2012, at *http://www.pasda.psu.edu/uci/Metadata-Display.aspx?entry=PASDA&file=PaStateRoads2009_01. xml&dataset=54.*

Pennsylvania Department of Transportation, 2009b, PennDOT —Pennsylvania municipality boundaries 2009: Pennsylvania Department of Transportation, Bureau of Planning and Research, Cartographic Information Division, digital dataset, accessed October 9, 2012, at *http://www.pasda.psu. edu/uci/MetadataDisplay.aspx?entry=PASDA&file=PaMun icipalities2009_01.xml&dataset=41.*

Pennsylvania Department of Transportation, 2009c, PennDOT —Pennsylvania county boundaries 2009: Pennsylvania Department of Transportation, Bureau of Planning and Research, Cartographic Information Division, digital dataset, accessed October 9, 2012, at *http://www.pasda.psu. edu/uci/MetadataDisplay.aspx?entry=PASDA&file=PaCou nty2009_01.xml&dataset=24.*

Rao, C.R., 1965, Linear statistical inference and its applications: New York, John Wiley and Sons, Inc., p. 435–513.

Rasmussen, P.P., Gray, J.R., Glysson, G.D., and Ziegler, A.C., 2009, Guidelines and procedures for computing time-series suspended-sediment concentrations and loads from in-stream turbidity-sensor and streamflow data: U.S. Geological Survey Techniques and Methods, book 3, chap. C4, 54 p. (Also available at *http://pubs.usgs.gov/tm/tm3c4/*.)

Sevon, W.D., 2000, Physiographic Provinces of Pennsylvania: Pennsylvania Geological Survey, 4th ser., Map 13, 1 sheet.

Simko Consulting, Inc., 2011, Draft Laurel Hill Creek water management plan: Eldorado Hills, Calif., Simko Consulting, Inc., 106 p.

Slattery, M.C., Walden, J., and Burt, T.P., 2000, Fingerprinting suspended sediment sources using mineral magnetic measurements—a quantitative approach, *in* Foster, I., ed., Tracers in geomorphology: New York, John Wiley and Sons, p. 309–322.

Sloto, R.A., and Olson, L.E., 2011, Estimated suspended-sediment loads and yields in the French and Brandywine Creek Basins, Chester County, Pennsylvania, water years 2008–09: U.S. Geological Survey Scientific Investigations Report 2011–5109, 31 p. (Also available at *http://pubs.usgs.gov/sir/2011/5109/support/sir2011-5109.pdf*.)

Somerset County Planning Commission, 2006, Somerset County comprehensive plan update, accessed February 22, 2012, at *http://www.co.somerset.pa.us/comprehensiveplan/plan_document.asp*.

Strahler, A.N., 1952, Dynamic basis of geomorphology: Geological Society of America Bulletin, v. 63, p. 923–938.

Taggart, J.E., Jr., ed., 2002, Analytical methods for chemical analysis of geologic and other materials: U.S. Geological Survey Open-File Report 02-223 [variously paged].

U.S. Department of Agriculture, 1987, Laurel Hill Creek Watershed, Somerset County, Pennsylvania watershed plan: U.S. Department of Agriculture, Soil Conservation Service Report, Somerset Conservation District, Harrisburg, Pa., unpublished report, chapters I–VIII.

U.S. Environmental Protection Agency, 1999, Guidance manual for compliance with the interim enhanced surface water treatment rule: EPA-815-R-99-010 [variously paginated].

U.S. Geological Survey, 2001, National land cover database 2001: U.S. Geological Survey digital dataset, accessed October 9, 2012, at *http://www.mrlc.gov/nlcd2001.php*.

U.S. Geological Survey, 2009, National hydrography dataset: digital dataset, accessed October 9, 2012, at *http://nhd.usgs.gov/data.html*.

Wainer, H., 1976, Robust statistics: a survey and some prescriptions: Journal of Educational Statistics, v. 1, no. 4, p. 285–312.

Walling, D.E., 2005, Tracing suspended sediment sources in catchments and river systems: Science of the Total Environment, v. 344, no. 1, p. 159–184.

Walling, D.E., and Woodward, J.C., 1992, Use of radiometric fingerprints to derive information on suspended sediment sources, in erosion and sediment transport monitoring programmes in river basins: Proceedings of the Oslo Symposium, August 1992, IAHS Publication Number 210, p. 153–164.

Appendixes 1, 2, 3, 4, and 5

Appendix 1. Results of field and laboratory analyses of water samples from Laurel Hill Creek near Bakersville, Pennsylvania (station 03079600).

Appendix 2. Results of field and laboratory analyses of water samples from Laurel Hill Creek at Ursina, Pennsylvania (station 03080000).

Appendix 3. Results of field and laboratory analyses of water samples from Laurel Hill Creek below Laurel Hill Creek Lake at Trent, Pennsylvania (station 03079655).

Appendix 4. Results of replicates and splits for analyzed tracers in samples collected from potential sediment sources, Laurel Hill Creek watershed, Somerset County, Pennsylvania. (Excel spreadsheet available online at *http://pubs.usgs.gov/sir/2012/5250/*)

Appendix 5. Laboratory data on tracers used in the sediment-source analysis, Laurel Hill Creek watershed, Somerset County, Pennsylvania. (Excel spreadsheet available online at *http://pubs.usgs.gov/sir/2012/5250/*)

Appendix 1.   Results of field and laboratory analyses of water samples from Laurel Hill Creek near Bakersville, Pennsylvania (station 03079600).

[ft³/s, cubic feet per second; mg/L, milligrams per liter; µS/cm, microsiemens per centimeter at 25°Celcius; °C, degrees Celcius; <, less than; E, estimated value; --, no data]

Sample start date	Sample start time	Sample end date	Sample end time	Sample type	Instantaneous discharge (ft³/s)	Mean discharge for composite sample (ft³/s)	Dissolved oxygen (mg/L)	pH, field (standard units)	Specific conductance, field (µS/cm)	Temperature (°C)	Chloride, dissolved (mg/L)	Ammonia, dissolved (mg/L)	Nitrate plus nitrite, dissolved (mg/L)	Nitrite, dissolved (mg/L)	Phosphorus, dissolved (mg/L)	Phosphorus, total (mg/L)	Total nitrogen, dissolved (mg/L)	Total nitrogen, total (mg/L)	Barium, dissolved (mg/L)	Suspended sediment (mg/L)	Suspended solids (mg/L)
7/28/09	1215	--	--	base flow	7.8	--	8.4	7.6	282	20.4	--	<.02	.8	.002	<.02	<.02	.9	.98	--	3	--
9/3/2009	1200	--	--	base flow	5.8	--	9.8	7.5	283	15.9	--	<.02	.75	.003	--	.01 E	1.04	.99	--	3	--
10/7/2009	1050	--	--	base flow	9.7	--	10.1	7.6	353	11.3	--	<.02	.72	.002 E	<.02	<.02	.92	.95	--	5	--
11/9/2009	1130	--	--	base flow	38	--	11.7	6.9	187	7.6	--	.017 E	.72	.002	<.02	<.02	.84	.86	--	23	--
12/17/2009	1030	--	--	storm	165	--	14.6	6.7	162	13	--	.015 E	.95	.002 E	<.02	<.02	1.1	1.05	--	4	--
12/17/2009	1031	--	--	duplicate	165	--	14.6	6.7	162	13	--	.015 E	.94	.002 E	<.02	<.02	1.03	1.09	--	--	--
12/17/2009	1115	--	--	storm	388	--	14	6.8	219	16	--	.014 E	1.16	.003	<.02	.05	1.25	1.4	--	32	--
3/11/2010	1130	--	--	storm	169	--	14.2	6.5	762	16	218	.051	.97	.004	<.02	.02 E	1.17	1.23	61.6	17	--
3/12/2010	800	3/12/2010	1105	composite	--	624	--	--	--	--	--	.046	.94	.004	<.02	.07	1.11	1.47	--	--	68
4/8/2010	1030	--	--	base flow	63	--	--	--	--	--	--	--	--	--	--	--	--	--	--	3	--
4/8/2010	1100	--	--	duplicate	61	--	--	--	--	--	--	--	--	--	--	--	--	--	--	14	--
4/19/2010	1320	--	--	base flow	67	--	11	7.2	223	8.7	--	.128	.59	.011	<.02	<.02	.99	.74	--	3	--
5/5/2010	1130	--	--	base flow	139	--	--	--	--	--	--	--	--	--	--	--	--	--	--	6	--
5/17/2010	1630	5/18/2010	1630	composite	--	368	--	--	--	--	--	--	--	--	--	--	--	--	--	--	58
6/7/2010	1240	--	--	base flow	33	--	8.4	7.4	275	17.7	--	.023	.69	.004	<.02	.06	.93	1.2	--	5	--
7/22/2010	1050	--	--	base flow	7.2	--	7.3	7.3	371	22.4	--	.036	.63	.004	<.02	<.02	.86	.9	--	2	--
8/31/2010	1000	--	--	base flow	5	--	7.9	7.6	421	18.2	--	.018 E	.6	.004	<.02	<.02	.81	.8	--	1	--
9/30/2010	1230	10/1/2010	1230	composite	--	87	--	--	--	--	--	.011 E	.5	.001 E	<.02	.06	.65	.67	--	--	48
10/5/2010	1030	--	--	base flow	8.9	--	9.8	7.4	345	10.4	--	.021	.39	.002 E	<.02	.06	.64	.93	--	1	--
10/19/2010	1230	10/19/2010	2030	composite	--	31	--	--	--	--	--	<.01	.99	.005	<.01	.01	1.32	1.3	--	--	6
10/22/2010	845	--	--	base flow	11	--	--	--	--	--	--	<.05	.37	.002	<.01	.01	.52	.57	--	3	--
11/16/2010	930	--	--	base flow	18	--	11.6	6.7	324	6.1	--	.015	.5	.006	<.01	<.01	.61	.63	--	3	--
11/16/2010	2030	11/16/2010	2030	composite	--	87	--	--	--	--	--	.024	.5	.003	.01	.03	.67	.88	--	--	20
3/24/2011	800	--	--	storm	254	--	--	--	--	--	--	--	--	--	--	--	--	--	--	124	--
3/24/2011	900	--	--	storm	254	--	--	--	--	--	--	--	--	--	--	--	--	--	--	40	--
3/24/2011	1000	--	--	storm	254	--	--	--	--	--	--	--	--	--	--	--	--	--	--	32	--
4/5/2011	430	4/5/2011	2030	composite	--	500	--	--	--	--	--	.038	.77	.003	<.01	.11	1	1.43	--	--	163
4/5/2011	900	--	--	storm	465	--	--	--	--	--	--	--	--	--	--	--	--	--	--	165	--
4/5/2011	1100	--	--	storm	515	--	--	--	--	--	--	--	--	--	--	--	--	--	--	162	--
4/5/2011	1300	--	--	storm	541	--	--	--	--	--	--	--	--	--	--	--	--	--	--	143	--

Appendix 1. Results of field and laboratory analyses of water samples from Laurel Hill Creek near Bakersville, Pennsylvania (station 03079600).—Continued

[ft³/s, cubic feet per second; mg/L, milligrams per liter; µS/cm, microsiemens per centimeter at 25°Celcius; °C, degrees Celcius; <, less than, E, estimated value; --, no data]

Sample start date	Sample end date	Sample end time	Sample start time	Sample type	Instantaneous discharge (ft³/s)	Mean discharge for composite sample (ft³/s)	Dissolved oxygen (mg/L)	pH, field (standard units)	Specific conductance, field (µS/cm)	Temperature (°C)	Chloride, dissolved (mg/L)	Ammonia, dissolved (mg/L)	Nitrate plus nitrite, dissolved (mg/L)	Nitrite, dissolved (mg/L)	Phosphorus, dissolved (mg/L)	Phosphorus, total (mg/L)	Total nitrogen, dissolved (mg/L)	Total nitrogen, total (mg/L)	Barium, dissolved (mg/L)	Suspended sediment (mg/L)	Suspended solids (mg/L)
4/5/2011	--	--	1700	storm	552	--	--	--	--	--	--	--	--	--	--	--	--	--	--	99	--
4/7/2011	--	--	1000	storm	218	--	12.6	5.7	201	6.4	--	0.012	0.97	0.002	<0.01	0.01	1.1	1.13	--	8	--
4/19/2011	4/20/2011	930	1930	composite	--	360	--	--	--	--	--	0.039	0.71	0.003	<0.01	0.03	0.9	1.09	--	--	55
4/19/2011	--	--	2100	storm	355	--	--	--	--	--	--	--	--	--	--	--	--	--	--	59	--
4/20/2011	--	--	200	storm	359	--	--	--	--	--	--	--	--	--	--	--	--	--	--	45	--
4/20/2011	--	--	700	storm	351	--	--	--	--	--	--	--	--	--	--	--	--	--	--	44	--
5/18/2011	5/18/2011	2330	930	composite	--	463	--	--	--	--	--	<0.01	0.72	0.001	<0.01	0.09	0.92	1.23	--	--	210
5/18/2011	--	--	1100	storm	395	--	--	--	--	--	--	--	--	--	--	--	--	--	--	208	--
5/18/2011	--	--	1200	storm	432	--	--	--	--	--	--	--	--	--	--	--	--	--	--	196	--
5/18/2011	--	--	1400	storm	446	--	--	--	--	--	--	--	--	--	--	--	--	--	--	162	--
5/18/2011	--	--	1700	storm	485	--	--	--	--	--	--	--	--	--	--	--	--	--	--	138	--
5/18/2011	--	--	1800	storm	531	--	--	--	--	--	--	--	--	--	--	--	--	--	--	170	--
5/18/2011	--	--	2000	storm	546	--	--	--	--	--	--	--	--	--	--	--	--	--	--	144	--
5/18/2011	--	--	2200	storm	578	--	--	--	--	--	--	--	--	--	--	--	--	--	--	106	--
6/20/2011	6/20/2011	1630	430	composite	--	176	--	--	--	--	--	--	--	--	--	--	--	--	--	--	108
6/20/2011	--	--	700	storm	175	--	--	--	--	--	--	--	--	--	--	--	--	--	--	103	--
6/20/2011	--	--	900	storm	208	--	--	--	--	--	--	--	--	--	--	--	--	--	--	106	--
6/20/2011	--	--	1000	storm	201	--	--	--	--	--	--	--	--	--	--	--	--	--	--	87	--
6/20/2011	--	--	1200	storm	188	--	--	--	--	--	--	--	--	--	--	--	--	--	--	61	--
6/20/2011	--	--	1500	storm	178	--	--	--	--	--	--	--	--	--	--	--	--	--	--	38	--
7/7/2011	--	--	940	base flow	11	--	--	--	344	20.8	--	--	--	--	--	--	--	--	--	--	--

Appendix 2.   Results of field and laboratory analyses of water samples from Laurel Hill Creek at Ursina, Pennsylvania (station 03080000).

[USGS, U.S. Geological Survey; PADEP, Pennsylvania Department of Environmenatal Protection; ft³/s, cubic feet per second; mg/L, milligrams per liter; µS/cm, microsiemens per centimeter at 25 C; C, degrees Celcius; N, nitrogen; P, phosphorus; CaCO₃, calcium carbonate; mL, milliliters; µg/L, micrograms per liter; <, less than; E, estimated value; M, not detected; --, no data]

Sample start date date	Sample start time	Sample end date	Sample end time	Analyzing laboratory	Instan-taneous discharge (ft³/s)	Mean dis-charge for composite sample (ft³/s)	Dissolved oxygen (mg/L)	pH, field (standard units)	Specific conduc-tance, field (µS/cm)	Tem-perature (°C)	Biochemi-cal oxygen demand, unfiltered (5 days at 20 °C, mg/L)	Dissolved solids (mg/L)	Hardness (mg/L as CaCo₃)
7/28/2009	900	--	--	USGS	25	--	8.4	6.6	193	20.4	--	--	--
9/3/2009	900	--	--	USGS	21	--	10	7	204	14.2	--	--	--
10/7/2009	900	--	--	USGS	32	--	10.4	7.4	179	12.1	--	--	--
11/9/2009	900	--	--	USGS	110	--	12.6	6.4	117	6.4	--	--	--
12/17/2009	930	--	--	USGS	429	--	15.1	6.8	110	0.9	--	--	--
1/27/2010	1000	--	--	USGS	1140	--	14.6	6.5	135	1.6	--	--	--
3/11/2010	940	--	--	USGS	650	--	14.7	6.1	358	1.2	--	--	--
4/20/2010	730	--	--	USGS	169	--	10.9	7	134	8.1	--	--	--
6/7/2010	915	--	--	USGS	85	--	9.1	7.4	146	18	--	--	--
7/22/2010	905	--	--	USGS	37	--	8.2	6.6	178	22.9	--	--	--
8/31/2010	845	--	--	USGS	6.9	--	8.1	7	262	19.6	--	--	--
9/30/2010	1430	10/1/2010	1430	USGS	--	227	--	7	149	--	--	--	--
10/5/2010	850	--	--	USGS	23	--	10.1	7	256	10.5	--	--	--
10/28/2010	1330	--	--	PADEP	57	--	10.6	7.7	185	13.1	1.1	122	41
11/17/2010	1300	--	--	PADEP	217	--	11.8	7.4	119	7.6	2	78	27
12/16/2010	1000	--	--	PADEP	243	--	13.9	7.8	182	0.1	0.9	112	30
1/19/2011	1300	--	--	PADEP	381	--	13.8	7.5	292	0.2	1.1	166	35
2/14/2011	1430	--	--	PADEP	339	--	13.9	7	220	0.7	1.3	108	30
3/15/2011	1300	--	--	PADEP	527	--	13.5	7	119	4.3	1	80	22
4/7/2011	1100	--	--	USGS	713	--	12.8	5.7	122	6.5	--	--	--
4/19/2011	1215	--	--	PADEP	598	--	10.7	7.4	112	9.4	1.4	62	24
5/18/2011	1145	--	--	PADEP	1690	--	10.4	7	75	12.9	1.2	62	20
6/7/2011	1345	--	--	PADEP	62	--	9.1	7.2	137	18	1.9	98	30
7/21/2011	1015	--	--	PADEP	21	--	8.1	7.9	209	26.4	0.7	122	43
8/10/2011	1000	--	--	PADEP	228	--	8.9	7.7	166	21.2	0.8	94	35
9/14/2011	1415	--	--	PADEP	143	--	9.2	7.7	123	20.6	0.3	84	29

Appendix 2. Results of field and laboratory analyses of water samples from Laurel Hill Creek at Ursina, Pennsylvania (station 03080000).—Continued

[USGS, U.S. Geological Survey; PADEP, Pennsylvania Department of Environmenatal Protection; ft³/s, cubic feet per second; mg/L, milligrams per liter; μS/cm, microsiemens per centimeter at 25 C; C, degrees Celcius; N, nitrogen; P, phosphorus; CaCO₃, calcium carbonate; mL, milliliters; μg/L, micrograms per liter; <, less than; E, estimated value; M, not detected; --, no data]

Sample start date date	Sample start time	Sample end date	Sample end time	Analyzing laboratory	Suspended solids, total (mg/L)	Calcium, total (mg/L)	Magnesium, total (mg/L)	Sodium, total (mg/L)	Acid neutralizing capacity, lab (mg/L as CaCO₃)	Bromide, dissolved (mg/L)	Chloride, dissolved (mg/L)	Sulfate, dissolved (mg/L)	Ammonia, dissolved (mg/L)
7/28/2009	900	--	--	USGS	--	--	--	--	--	--	--	--	<0.02
9/3/2009	900	--	--	USGS	--	--	--	--	--	--	--	--	<0.02
10/7/2009	900	--	--	USGS	--	--	--	--	--	--	--	--	<0.02
11/9/2009	900	--	--	USGS	--	--	--	--	--	--	--	--	0.011 E
12/17/2009	930	--	--	USGS	--	--	--	--	--	--	--	--	<0.02
1/27/2010	1000	--	--	USGS	--	--	--	--	--	--	--	--	0.021
3/11/2010	940	--	--	USGS	--	--	--	--	--	--	92.9	--	<0.02
4/20/2010	730	--	--	USGS	--	--	--	--	--	--	--	--	0.069
6/7/2010	915	--	--	USGS	--	--	--	--	--	--	--	--	0.013 E
7/22/2010	905	--	--	USGS	--	--	--	--	--	--	--	--	0.011 E
8/31/2010	845	--	--	USGS	--	--	--	--	--	--	--	--	0.013 E
9/30/2010	1430	10/1/2010	1430	USGS	--	--	--	--	--	--	--	--	<0.02
10/5/2010	850	--	--	USGS	--	--	--	--	--	--	--	--	<0.01
10/28/2010	1330	--	--	PADEP	<5	12.5	2.4	21.2	23	<0.1	40.2	11.6	--
11/17/2010	1300	--	--	PADEP	<5	8.1	1.7	10.5	13	<0.1	19.1	10.2	--
12/16/2010	1000	--	--	PADEP	<5	8.6	2.1	18.8	10	<0.1	36.8	11.3	--
1/19/2011	1300	--	--	PADEP	<5	10.4	2.2	39	10	<0.1	67.9	10.3	--
2/14/2011	1430	--	--	PADEP	28	9	1.9	29.2	9	<0.1	49.2	9.3	--
3/15/2011	1300	--	--	PADEP	<5	6.5	1.4	12.5	7	<0.1	22.6	9.5	--
4/7/2011	1100	--	--	USGS	--	--	--	--	--	--	--	--	<0.01
4/19/2011	1215	--	--	PADEP	8	7	1.5	11.2	12	<0.1	19.6	9.6	--
5/18/2011	1145	--	--	PADEP	28	5.7	1.3	6.1	8	<0.1	9.3	10.2	--
6/7/2011	1345	--	--	PADEP	6	9	1.8	12.5	15	<0.1	22.9	10.1	--
7/21/2011	1015	--	--	PADEP	<5	13.1	2.4	21.4	25	<0.1	42	9.7	--
8/10/2011	1000	--	--	PADEP	8	10.9	1.8	18.8	22	<0.1	32	8.6	--
9/14/2011	1415	--	--	PADEP	16	8.9	1.6	12.6	15	<0.1	19.4	9.3	--

Appendix 2. Results of field and laboratory analyses of water samples from Laurel Hill Creek at Ursina, Pennsylvania (station 03080000).—Continued

[USGS, U.S. Geological Survey; PADEP, Pennsylvania Department of Environmental Protection; ft³/s, cubic feet per second; mg/L, milligrams per liter; µS/cm, microsiemens per centimeter at 25 C; C, degrees Celcius; N, nitrogen; P, phosphorus; CaCO₃, calcium carbonate; mL, milliliters; µg/L, micrograms per liter; <, less than; E, estimated value; M, not detected; --, no data]

Sample start date date	Sample start time	Sample end date	Sample end time	Analyzing laboratory	Ammonia, total (mg/L as N)	Nitrate plus nitrite, dissolved (mg/L as N)	Nitrate, dissolved (mg/L as N)	Nitrite, dissolved (mg/L as N)	Nitrite, total (mg/L as N)	Orthophosphate, total (mg/L as P)	Phosphorus, dissolved (mg/L)	Phosphorus, total (mg/L)	Total nitrogen, dissolved (mg/L)
7/28/2009	900	--	--	USGS	--	0.24	--	0.001 E	--	--	<0.02	<0.02	0.37
9/3/2009	900	--	--	USGS	--	0.17	--	<0.002	--	--	--	<0.02	0.27
10/7/2009	900	--	--	USGS	--	0.17	--	<0.002	--	--	<0.02	<0.02	0.28
11/9/2009	900	--	--	USGS	--	0.51	--	<0.002	--	--	<0.02	<0.02	0.6
12/17/2009	930	--	--	USGS	--	0.66	--	<0.002	--	--	<0.02	<0.02	0.7
1/27/2010	1000	--	--	USGS	--	0.87	--	0.001 E	--	--	<0.02	0.02 E	0.97
3/11/2010	940	--	--	USGS	--	0.77	--	0.002 E	--	--	<0.02	<0.02	0.9
4/20/2010	730	--	--	USGS	--	0.37	--	0.007	--	--	<0.02	<0.02	0.58
6/7/2010	915	--	--	USGS	--	0.47	--	0.003	--	--	<0.02	<0.02	0.61
7/22/2010	905	--	--	USGS	--	0.16	--	0.002	--	--	<0.02	<0.02	0.25
8/31/2010	845	--	--	USGS	--	0.08	--	0.001 E	--	--	<0.02	0.02 E	0.21
9/30/2010	1430	10/1/2010	1430	USGS	--	0.33	--	0.002 E	--	--	<0.02	0.24	0.52
10/5/2010	850	--	--	USGS	--	0.19	--	0.001	--	--	<0.01	<0.01	0.29
10/28/2010	1330	--	--	PADEP	<0.02	--	0.06	--	<0.04	<0.01	--	<0.01	0.12
11/17/2010	1300	--	--	PADEP	<0.02	--	0.33	--	<0.04	0.02	--	0.012	0.48
12/16/2010	1000	--	--	PADEP	<0.02	--	0.97	--	<0.04	0.01	--	<0.01	1.1
1/19/2011	1300	--	--	PADEP	<0.02	--	0.74	--	<0.04	<0.01	--	<0.01	0.81
2/14/2011	1430	--	--	PADEP	<0.02	--	0.88	--	<0.04	<0.01	--	<0.01	1
3/15/2011	1300	--	--	PADEP	<0.02	--	0.74	--	<0.04	<0.01	--	<0.01	0.71
4/7/2011	1100	--	--	USGS	--	0.72	--	0.001	--	--	<0.01	0.01	0.82
4/19/2011	1215	--	--	PADEP	<0.02	--	0.61	--	<0.04	0.01	--	0.01	0.71
5/18/2011	1145	--	--	PADEP	0.04	--	0.5	--	<0.04	<0.01	--	0.032	0.7
6/7/2011	1345	--	--	PADEP	<0.02	--	0.34	--	<0.04	<0.01	--	<0.01	0.35
7/21/2011	1015	--	--	PADEP	0.04	--	0.19	--	<0.04	<0.01	--	0.01	0.32
8/10/2011	1000	--	--	PADEP	0.04	--	0.3	--	<0.04	<0.01	--	0.013	0.54
9/14/2011	1415	--	--	PADEP	<0.02	--	0.56	--	<0.04	<0.01	--	<0.01	0.73

Appendix 2.   Results of field and laboratory analyses of water samples from Laurel Hill Creek at Ursina, Pennsylvania (station 03080000).—Continued

[USGS, U.S. Geological Survey; PADEP, Pennsylvania Department of Environmenatal Protection; ft³/s, cubic feet per second; mg/L, milligrams per liter; µS/cm, microsiemens per centimeter at 25 C; C, degrees Celcius; N, nitrogen; P, phosphorus; CaCO₃, calcium carbonate; mL, milliliters; µg/L, micrograms per liter; <, less than; E, estimated value; M, not detected; --, no data]

Sample start date	Sample start time	Sample end date	Sample end time	Analyzing laboratory	Total nitrogen, total (mg/L)	Fecal coliform (colonies per 100 mL)	Aluminum, dissolved (µg/L)	Aluminum, total (µg/L)	Barium, dissolved (µg/L)	Cadmium, dissolved (µg/L)	Copper, dissolved (µg/L)	Copper, total (µg/L)	Iron, dissolved (µg/L)
7/28/2009	900	--	--	USGS	0.39	--	--	--	--	--	--	--	--
9/3/2009	900	--	--	USGS	0.65	--	--	--	--	--	--	--	--
10/7/2009	900	--	--	USGS	0.3	--	--	--	--	--	--	--	--
11/9/2009	900	--	--	USGS	0.62	--	--	--	--	--	--	--	--
12/17/2009	930	--	--	USGS	0.8	--	--	--	--	--	--	--	--
1/27/2010	1000	--	--	USGS	1.04	--	--	--	--	--	--	--	--
3/11/2010	940	--	--	USGS	0.92	--	--	--	49.7	--	--	--	--
4/20/2010	730	--	--	USGS	0.51	--	--	--	--	--	--	--	--
6/7/2010	915	--	--	USGS	0.65	--	--	--	--	--	--	--	--
7/22/2010	905	--	--	USGS	0.28	--	--	--	--	--	--	--	--
8/31/2010	845	--	--	USGS	0.25	--	--	--	--	--	--	--	--
9/30/2010	1430	10/1/2010	1430	USGS	0.97	--	--	--	--	--	--	--	--
10/5/2010	850	--	--	USGS	0.33	--	--	--	--	--	--	--	--
10/28/2010	1330	--	--	PADEP	--	--	11	M	--	<0.2	<4	<4	<20
11/17/2010	1300	--	--	PADEP	--	90	22	100	--	<0.2	<4	<4	50
12/16/2010	1000	--	--	PADEP	--	--	43	M	--	<0.2	<4	<4	50
1/19/2011	1300	--	--	PADEP	--	<10	14	M	--	<0.2	<4	<4	30
2/14/2011	1430	--	--	PADEP	--	10	19	M	--	<0.2	<4	<4	30
3/15/2011	1300	--	--	PADEP	--	10	26	100	--	<0.2	<4	<4	<20
4/7/2011	1100	--	--	USGS	0.87	--	--	--	--	<0.2	<4	--	--
4/19/2011	1215	--	--	PADEP	--	20	24	100	--	<0.2	<4	<4	<20
5/18/2011	1145	--	--	PADEP	--	500	59	500	--	<0.2	<4	<4	80
6/7/2011	1345	--	--	PADEP	--	140	18	M	--	<0.2	<4	<4	30
7/21/2011	1015	--	--	PADEP	--	--	22	M	--	<0.2	<4	<4	30
8/10/2011	1000	--	--	PADEP	--	190	28	200	--	<0.2	<4	<4	60
9/14/2011	1415	--	--	PADEP	--	40	31	M	--	<0.2	<4	<4	40

Appendix 2. Results of field and laboratory analyses of water samples from Laurel Hill Creek at Ursina, Pennsylvania (station 03080000).—Continued

[USGS, U.S. Geological Survey; PADEP, Pennsylvania Department of Environmental Protection; ft³/s, cubic feet per second; mg/L, milligrams per liter; µS/cm, microsiemens per centimeter at 25 C; C, degrees Celcius; N, nitrogen; P, phosphorus; CaCO₃, calcium carbonate; mL, milliliters; µg/L, micrograms per liter; <, less than, E, estimated value; M, not detected; --, no data]

Sample start date date	Sample start time	Sample end date	Sample end time	Analyzing laboratory	Iron, total (µg/L)	Lead, dissolved (µg/L)	Lead, total (µg/L)	Manganese, dissolved (µg/L)	Manganese, total (µg/L)	Nickel, dissolved (µg/L)	Nickel, total (µg/L)	Strontium, total (µg/L)	Zinc, dissolved (µg/L)
7/28/2009	900	--	--	USGS	--	--	--	--	--	--	--	--	--
9/3/2009	900	--	--	USGS	--	--	--	--	--	--	--	--	--
10/7/2009	900	--	--	USGS	--	--	--	--	--	--	--	--	--
11/9/2009	900	--	--	USGS	--	--	--	--	--	--	--	--	--
12/17/2009	930	--	--	USGS	--	--	--	--	--	--	--	--	--
1/27/2010	1000	--	--	USGS	--	--	--	--	--	--	--	--	--
3/11/2010	940	--	--	USGS	--	--	--	--	--	--	--	--	--
4/20/2010	730	--	--	USGS	--	--	--	--	--	--	--	--	--
6/7/2010	915	--	--	USGS	--	--	--	--	--	--	--	--	--
7/22/2010	905	--	--	USGS	--	--	--	--	--	--	--	--	--
8/31/2010	845	--	--	USGS	--	--	--	--	--	--	--	--	--
9/30/2010	1430	10/1/2010	1430	USGS	--	--	--	--	--	--	--	--	--
10/5/2010	850	--	--	USGS	--	--	--	--	--	--	--	--	--
10/28/2010	1330	--	--	PADEP	50	<1	<1	M	M	<4	<4	40	--
11/17/2010	1300	--	--	PADEP	260	<1	<1	M	50	<4	<4	30	--
12/16/2010	1000	--	--	PADEP	120	<1	<1	20	20	<4	<4	40	--
1/19/2011	1300	--	--	PADEP	80	<1	<1	M	10	<4	<4	40	--
2/14/2011	1430	--	--	PADEP	110	<1	<1	M	20	<4	<4	30	M
3/15/2011	1300	--	--	PADEP	100	<1	<1	10	20	<4	<4	20	M
4/7/2011	1100	--	--	USGS	--	--	--	--	--	--	--	--	--
4/19/2011	1215	--	--	PADEP	180	<1	<1	10	20	<4	<4	20	M
5/18/2011	1145	--	--	PADEP	1020	<1	<1	20	90	<4	<4	20	M
6/7/2011	1345	--	--	PADEP	60	<1	<1	M	10	<4	<4	30	<5
7/21/2011	1015	--	--	PADEP	60	<1	<1	30	40	M	M	50	<5
8/10/2011	1000	--	--	PADEP	400	<1	<1	10	60	<4	<4	30	M
9/14/2011	1415	--	--	PADEP	150	<1	<1	M	10	<4	<4	30	<5

Appendix 2. Results of field and laboratory analyses of water samples from Laurel Hill Creek at Ursina, Pennsylvania (station 03080000).—Continued

[USGS, U.S. Geological Survey; PADEP, Pennsylvania Department of Environmenatal Protection; ft³/s, cubic feet per second; mg/L, milligrams per liter; µS/cm, microsiemens per centimeter at 25 C; C, degrees Celcius; N, nitrogen; P, phosphorus; CaCO₃, calcium carbonate; mL, milliliters; µg/L, micrograms per liter; <, less than; E, estimated value; M, not detected; --, no data]

Sample start date date	Sample start time	Sample end date	Sample end time	Analyzing laboratory	Zinc, total (µg/L)	Arsenic, dissolved (µg/L)	Boron, total (µg/L)	Selenium, dissolved (µg/L)	Suspended sediment (mg/L)
7/28/2009	900	--	--	USGS	--	--	--	--	2
9/3/2009	900	--	--	USGS	--	--	--	--	<0.5
10/7/2009	900	--	--	USGS	--	--	--	--	<0.5
11/9/2009	900	--	--	USGS	--	--	--	--	<0.5
12/17/2009	930	--	--	USGS	--	--	--	--	2
1/27/2010	1000	--	--	USGS	--	--	--	--	22
3/11/2010	940	--	--	USGS	--	--	--	--	10
4/20/2010	730	--	--	USGS	--	--	--	--	3
6/7/2010	915	--	--	USGS	--	--	--	--	2
7/22/2010	905	--	--	USGS	--	--	--	--	2
8/31/2010	845	--	--	USGS	--	--	--	--	0.5
9/30/2010	1430	10/1/2010	1430	USGS	--	--	--	--	64
10/5/2010	850	--	--	USGS	--	--	--	--	<0.5
10/28/2010	1330	--	--	PADEP	--	<3	<200	<7	--
11/17/2010	1300	--	--	PADEP	--	<3	<200	<7	--
12/16/2010	1000	--	--	PADEP	--	<3	<200	<7	--
1/19/2011	1300	--	--	PADEP	--	<3	<200	<7	--
2/14/2011	1430	--	--	PADEP	10	<3	<200	<7	--
3/15/2011	1300	--	--	PADEP	M	<3	<200	<7	--
4/7/2011	1100	--	--	USGS	--	--	--	--	6
4/19/2011	1215	--	--	PADEP	M	<3	<200	<7	--
5/18/2011	1145	--	--	PADEP	20	<3	<200	<7	--
6/7/2011	1345	--	--	PADEP	<5	<3	<200	<7	--
7/21/2011	1015	--	--	PADEP	<5	<3	<200	<7	--
8/10/2011	1000	--	--	PADEP	10	<3	<200	<7	--
9/14/2011	1415	--	--	PADEP	<5	<3	<200	<7	--

Appendix 3. Results of field and laboratory analyses of water samples from Laurel Hill Creek below Laurel Hill Lake at Trent, Pennsylvania (station 03079655).

[ft³/s, cubic feet per second; mg/L, milligrams per liter; µS/cm, microsiemens per centimeter at 25°Celcius; °C, degrees Celcius; <, less than; --, no data]

Sample date	Sample time	Instan- taneous discharge (ft³/s)	Dissolved oxygen (mg/L)	pH, field (standard units)	Specific conduc- tance, field (µS/cm)	Temper- ature (°C)	Ammonia, dissolved (mg/L)	Nitrate plus nitrite, dissolved (mg/L)	Nitrite, dissolved (mg/L)	Phos- phorus, dissolved (mg/L)	Phos- phorus, total (mg/L)	Total nitrogen, dissolved (mg/L)	Total nitrogen, total (mg/L)	Suspended sediment concen- tration (mg/L)
7/28/09	1100	13	7.9	7.3	227	21.3	0.016	0.53	0.004	<0.02	<0.02	0.68	0.71	4
9/3/2009	1045	8.6	8.5	7.1	275	17.7	0.021	0.31	0.003	--	<0.02	0.51	0.49	2
6/7/2010	1045	39	8	7.2	197	19.2	0.037	0.56	0.005	<0.02	<0.02	0.75	0.82	2

www.ingramcontent.com/pod-product-compliance
Lightning Source LLC
Chambersburg PA
CBHW081621170526
45166CB00009B/3054